Creativity Theory and Action in Education

Volume 7

Educational settings represent sites of creative possibility. They also represent the manifestation of some of the most persistent and dogmatic beliefs about teaching and learning. This series aims to push the frontiers of creativity theory, research, and practice in educational settings. Specifically, this series endeavors to provide a venue for disseminating the kinds of provocative thinking and cutting-edge research that can promote more creative approaches to teaching and learning. The focus of the series is on mainstream (rather than gifted or other specialized) educational settings. Another aspect worthy of exploration is domain specific or domain general view of creativity- one that has hitherto been the speculation of cognitive science but one that can be brought to the forefront of existing treatments of creativity. A final (and general) area of investigation is artistic, ecological, cultural and anthropological aspects of creativity that have been ignored by the community.

This Series:

• Capitalizes on the growing international interest and concern about the breakdown of creativity in everyday schools and classrooms

• Provides fresh thinking on complex issues and challenges pertaining to theory and practice aimed at promoting creativity in educational settings

• Challenges existing dogmas and overly narrow conceptions of teaching, learning, and creativity (e.g., creativity being separated from academic learning and linked to gifted education)

• Spotlights new theories, methodologies and approaches to studying and enacting creativity in a variety of domains, contexts, and levels (early childhood through higher education)

The Editors of this Series welcome proposals for edited and authored volumes that provide provocative and original explorations of creative theory, methodology and action in educational settings. This includes international and multidisciplinary perspectives on creativity across and within K-12, university, online and informal educational settings (e.g., museums, organizations, clubs, and groups). The audience for this series includes creativity and educational researchers, graduate students, practicing educators, and educational thought leaders.

Proposal Review Process:

All proposals are reviewed by series and acquisition editors. There are two types of review: *expedited* and *extended.* Expedited reviews occur in cases where the series and acquisition editors have determined that a project should or should not move forward based on the fit and potential of proposal to contribute to the series. Extended reviews occur in cases where proposals take on more specialized topics and would benefit from additional input provided by experts serving on the series advisory board and, in some cases, relevant external reviewers who can make a recommendation about the potential merit of a particular project. Extended reviews may also include an exit review of the completed work by board members or external reviewers.

Danah Henriksen • Punya Mishra
Editors

Creative Provocations: Speculations on the Future of Creativity, Technology & Learning

Springer

Editors
Danah Henriksen
Mary Lou Fulton Teachers College
Arizona State University
Tempe, AZ, USA

Punya Mishra
Mary Lou Fulton Teachers College
Arizona State University
Tempe, AZ, USA

ISSN 2509-5781 ISSN 2509-579X (electronic)
Creativity Theory and Action in Education
ISBN 978-3-031-14548-3 ISBN 978-3-031-14549-0 (eBook)
https://doi.org/10.1007/978-3-031-14549-0

This Springer imprint is published by the registered company Springer Nature Switzerland AG
The registered company address is: Gewerbestrasse 11, 6330 Cham, Switzerland

To current and past members of the Deep Play Research Group. Thank you for being part of this decade-long journey through creativity, technology, and learning. Here's to 10 more years and beyond.

Preface

ποταμοῖσι τοῖσιν αὐτοῖσιν ἐμβαίνουσιν, ἕτερα καί ἕτερα ὕδατα ἐπιρρεῖ — Ἡράκλειτος
(*Ever-newer waters flow on those who step into the same rivers* — Heraclitus)

If you do not change direction, you may end up where you are heading — Lao Tzu

It is questions with no answers that set the limit of human possibilities, describe the boundaries of human existence — Milan Kundera

We live in times of exponential change—fueled in large part by technological innovation and human creativity. These changes are dramatically reshaping how we live, work, think, create, and make sense of the world. Ideas and knowledge can be discovered, created, and shared faster and more widely than ever before. This has led to a flowering of creativity at the individual and social level in historically unprecedented ways. At the same time, it has also led to a plague of misinformation, tribalism, and new societal and individual problems. These forces combined with the forces of globalization and rampant growth, often in the face of significant inequalities (local and global), bring significant social and economic challenges. Ironically, the inequities created by advancing technological changes can in turn deprive many people of the technological, economic, and social benefits that the changes brought about.

It can be argued that, at some level, this has been a failure of education—that the challenges we face, whether global climate change or the spread of misinformation, social unrest, the rise of tribalism, among others, can be traced to failures within our educational systems. The COVID pandemic and the scattered, inconsistent, inequitable, and often ineffective response to it, as well as the range of educational innovations that emerged from the pandemic to, are cases in point. The pandemic itself is an example of one of many "planet-sized" (Gidley 2016) global disruptions, which will continue to stress our already inefficient political, economic, and cultural institutions. The accelerating influx of information and technology, in a volatile, uncertain, complex, and ambiguous, i.e., VUCA (Cousins 2018) environment, underlines the key role of creativity in education, since uncertainty and unanticipated situations and problems require creative thinking and novel solutions.

At the foundation of this book is an acknowledgment that the only way these current and upcoming challenges can be addressed at scale is through creativity and ingenuity: through the development of new sustainable, humane, and creative technological educational ecosystems. Only through education can these changes become part of the broader global social fabric to prepare us, individually and collectively, to address and design better and more equitable futures. In a world where change is the only constant, we must find ways to learn and educate, through creativity and with the technological tools at hand (and those emerging in the future), to determine our place in this ever-evolving, emergent future.

If the COVID crisis and its impact on education has demonstrated anything, it is that technology (particularly the Internet) has an important role to play in equitable (or inequitable) educational futures. Access to digital or Internet technologies may be the biggest equity issue facing education through the pandemic and beyond. Further, existing non-digital pedagogical practices do not necessarily map effortlessly onto the digital world, which means that creativity is needed in translating between digital and non-digital mediums effectively. Educational institutions and contexts have typically molded technology to fit existing education practices, rather than take advantage of the affordances of the technology to reimagine what education could be. While there are examples of creative uses of technology, they are often idiosyncratic, not systemic, and limited by preconceived structures and practices. For instance, during the pandemic, when schools or universities moved to remote settings, some individual educators demonstrated creativity (technologically and pedagogically) to connect with their students and continue their learning. However, within most systems it was unclear that any basic assumptions about education and the use of technology had been fundamentally questioned and reframed, and there was limited discussion of technology and its potential role in advancing creativity in learning contexts.

That said, incorporating creativity into education requires an accurate and grounded understanding and articulation of the construct itself. Too often, creativity in education is framed in ways that are out of step with global changes and with little understanding of the role of technology in the present and emergent futures. If creativity is to truly find its value in education, it must build new understandings and framings of the relationship between creativity and technology. For instance, the emergence and potential of artificial intelligence forces us to confront the possibility of creativity beyond or outside of the human. This is an idea that is rarely considered in popular societal discussions, and even less so in education—but it prompts a posthuman perspective on creativity that demands consideration. The same kind of (re)consideration can emerge from many different new, networked, and digital futures that are emerging and will continue to unfold.

This weakness in conceptualization and grounding of the meaning of creativity and the role of technology in fostering it in learning is in sharp contrast to developments outside of education, in society at large. The advent of digital tools and the Internet has unleashed a tidal wave of ground-up creativity, with people across the globe engaging in polished, socially savvy, and technologically mediated creative behaviors (through tools such as Tiktok or Twitter, Instagram, YouTube, and more).

While creative technology use in education broadly lags, both in quantitative terms (access and time spent) and qualitatively (what is done with digital tools and connectivity), learners (and society overall) have taken up new tools as opportunities to create, share, explore, construct, inquire, communicate, and express themselves—all of which are universally recognized as being good learning behaviors.

At the heart of this book is the idea that to foreground creativity within educational systems and take advantage of technological affordances, we must better understand both creativity and the inherent propensities of the technology to support and enhance it (while being sensitive to and aware of inequities that restrict access to resources for large swaths of people). We suggest that there are organic synergies between a societal or distributed view of creativity and the affordances of digital technologies to change how we think of both creativity and technology in educational situations.

It is also important to acknowledge that the affordances of technology to affect thinking, learning, or creativity did not originate with the Internet, computers, or digital media. Pea's (1987) definition of cognitive technologies as "any medium that helps transcend the limitations of the mind (e.g., attention to goals, short-term memory span) in thinking, learning, and problem-solving activities" (p. 91) includes systems of written language or mathematical notation. Every new technology, whether oral/written language, pencil and paper, the digital computer, or the Internet, has transformed how we learn and engage with the world. Thus, we do not assert that digital technologies are the only or the best path to creativity in education, rather, we highlight how digital technologies are central to our world and its future, and more importantly, to the world and future that learners inhabit. They provide powerful affordances for inquiry, information access, creating, editing, remixing, and sharing. However, these affordances offer little unless they are aligned with meaningful learning goals, creative purposes, student needs, or the cultural geography of the world and the classroom's place within it.

It is important for learners, educators, and creative users of tools to be aware of affordances and consider what available (digital or non-digital) resources have to offer. Technological tools are not just individual gadgets but are embedded in and constitute systems of knowledge, skills, and organization. Koehler et al. (2011) argue that the idea of repurposing, or melioration (Passig 2007), is at the heart of the creative uses of technology. Educators and learners need to develop fluency with the tools and recognize which rules to bend, which they can break, and which to leave as is (Koehler et al. 2011). This suggests that a technology's purpose, while shaped by affordances and properties, is not pre-determined. Rather, the inextricable interactions of objects, environment, and human psyche determine its potential and significance. Despite the obvious potentialities and dramatic shifts wrought by digital technologies, it is still an individual's (or collective's) evaluative ability and creative use of tools that dictate what they do or create.

In between the good, the bad, the unimagined, or the unintended consequences, there is a wide range of complex and intertwined issues, topics, arguments, and points of view to understanding the potential and possible futures of what it means to live and learn creatively with and through technologies (Henriksen et al. 2016).

These issues have resonance in education, as an arena deeply tied to the exploration, development, construction, or deconstruction of knowledge (Peppler et al. 2016) for the future.

This book explores the complex, yet critical, relationship between technology and creativity, specifically in educational contexts. Chapters in this volume, by some of the foremost thinkers and researchers in the areas of creativity and/or technology, examine the impact of recent and future developments in technology on creativity, teaching, and learning. These chapters, which emerged from authors with expertise across disciplines and with diverse perspectives, all consider, analyze, or speculate in ways that might inform the field of education.

Given the rapid and shifting challenges and opportunities before us, the best way to understand the future may be through *grounded imagination*—combining expertise with thoughtful speculation. In seeking chapters for this book, we emphasized that authors should feel free to use their expertise to speculate, to imagine, to step outside the standard academic boundaries, and if or where possible, to bring new perspectives on this rapidly evolving future—exploring issues, paradoxes, tensions, and points of interest around creativity and technology relevant to learning. Thematically speaking, the chapters in this book take on a range of perspectives, often in agreement and sometimes in opposition. That said, certain foundational ideas stand out.

First, education is always about preparing learners for the future, and the future will always be uncertain. It behooves us to ask, what are certain values-driven principles of creative learning under such uncertainty? To ask, how existing and new technologies might help or further impede efforts aimed at promoting creative learning in and beyond the walls of schools and classrooms?

Second, creativity is an essential skill and one that can be developed. That said, the meaning and role of creativity in our world is deeply connected to context. This means that creative education or education for creativity is denied to many (due to existing disparities within our broader societal contexts), so there is a profound issue of equity involved.

Third, it is useful to expand our understanding of creativity to recognize its social-cultural elements, embedded in contexts, and histories, which develops not just in individuals but in broader and larger "clusters" or communities of practice. And these clusters/communities are often connected through online methods of connection and mediation—which can have negative and positive consequences.

Fourth, technology has a complex relationship with creativity and education. Technologies offer significant possibilities but can also be distractive or inhibit creativity or get in the way of learning. The COVID pandemic in particular thrust educators into using technology for teaching and learning—leading to unexpected challenges and breakdowns, but also emergent possibilities, particularly in generative, creative, and embodied artistic spaces, such as therapy, music, and theater/dance. Technologies also mediate the creative process, but newer forms of technological development can be disruptive as well, at levels that may not have been possible earlier. In particular, artificial intelligence, and all the potential that is emerging in this field, could change our perceptions of the human relationship to creativity, leading to new visions of a posthuman future.

Finally, there are also significant challenges that creativity faces within broader political and educational systems. Although an emphasis on creativity can be generally seen as a positive thing, the manner in which it is embedded within broader social, economic, and educational systems is worthy of attention. This means we must continually question how creativity is defined, instantiated, and supported. Who is included in this conversation and how should we seek to measure or evaluate it? It requires us to question normative and foundational issues, because not doing so may lead to the perpetuation and crystallization of existing frameworks that solidify existing disparities.

The authors in this book are some of the most provocative and thoughtful scholars in the field, and in each of their pieces they explore key provocations, raise critical questions, argue for and against particular positions, and, just as importantly, speculate on potentials and possibilities. These chapters shed light on the role of creativity and our technological future for thinking and learning, across a range of educational settings. They individually and collectively highlight the complexities and paradoxes inherent in these issues, exploring them from multiple perspectives, while at the same time aiming to put a "stake in the ground" about their own perspective, and through that provide a range of "creative provocations" to help us think about these emergent futures.

Tempe, AZ, USA

Danah Henriksen
Punya Mishra

References

Cousins, B. (2018). Design thinking: Organizational learning in VUCA environments. *Academy of Strategic Management Journal, 17*(2), 1–18.

Gidley, J. (2016). *Postformal education: A philosophy for complex futures*. Switzerland: Springer.

Henriksen, D., Mishra, P., & Fisser, P. (2016). Infusing creativity and technology in 21st century education: A systemic view for change. *Educational Technology & Society, 19*(3), 27–37.

Koehler, M. J., Mishra, P., Bouck, E. C., DeSchryver, M., Kereluik, K., Shin, T. S., & Wolf, L. G. (2011). Deep-play: Developing TPACK for 21st century teachers. *International Journal of Learning Technology, 6*(2), 146–163.

Passig, D. (2007). Melioration as a higher thinking skill of future intelligence. *Teachers College Record, 109*(1), 24–50.

Pea, R. D. (1987). Cognitive technologies for mathematics education. In A. H. Schoenfeld (Ed.), *Cognitive science and mathematics education* (pp. 89–122). Hillsdale, N.J.: Lawrence Erlbaum Associates.

Peppler, K., Halverson, E., & Kafai, Y. B. (Eds.). (2016). *Makeology: Makerspaces as learning environments (Volume 1)* (Vol. 1). Routledge.

Acknowledgments

Any book such as this is the product of many individuals and groups working together. In this case, the complexity and challenges in the work were compounded by the presence of a global pandemic and the concomitant chaos is caused in each of our lives. The fact that the authors of the chapters still managed to bring their knowledge to bear productively and thoughtfully, working through many pandemic-related disruptions, is a testament to their expertise and commitment to knowledge advancement and production in this space.

So, our first order of thanks are to the individual authors (and author teams) who worked through a global pandemic and still managed to write and get this set of thoughtful and provocative chapters to us. We are deeply grateful. A special thanks also goes to Vlad Glaveanu, Giovanni Corazza, and Ingunn Ness—for taking on the task of reading through all the articles and coming up with what we believe is a creative and wonderful solution for our last chapter—a trialogue. The ideas in this book defy easy classification and we believe what they have done in the last chapter honors the richness and poly-vocality of the domain in ways a traditional chapter could not have.

We would also like to thank Ron Beghetto and Bharath Sriraman, the editors for the book series *Creativity Theory and Action in Education*, which this volume is part of, for reaching out to us several years ago. This book would not have seen the light of day without their faith in us being able to pull it off. Thanks also to the entire publishing team at Springer. Their regular emails prevented us from doing that most academic of tasks, procrastinating. We truly mean it when we say their subtle (and not so subtle) nudges kept us on task and ensured that this book would actually be completed.

We also thank the *Mary Lou Fulton Teachers College* at *Arizona State University,* not just for our paychecks (although that helps), but for creating a scholarly atmosphere that encourages and supports creative thinking, innovation, and academic rigor, as well as trusting scholars enough to allow for informed and imaginative speculative thought.

We would like to take a moment to thank each other, for shared accountability, a willingness to step up when other commitments got in the way. This is a genuine scholarly partnership and has been, as always, great fun.

Phoenix, AZ, USA
July 2022

Contents

Chapter 1
Blueprints for a Creativity Curriculum

Anthony Brandt

Abstract Decades of cognitive research have arrived at a consensus: all children have creative potential; creativity depends as much on exposure and opportunity as on hereditary factors; and it can be improved through instruction and practice. Yet cultivating creativity has increasingly become a matter of equity. As industrialized countries move towards greater automation and human labor shifts towards jobs requiring flexible thinking, millions of children are being disadvantaged. How to design a curricular framework at the necessary scale? Building it around divergent thinking has a host of advantages. The concepts are straight-forward and easy to define. They are well-established in the literature, making it easier to persuade school boards and state legislatures. They are trans-disciplinary, as applicable to STEM fields as to the arts. Successful exemplars exist, with objective ways to measure them. Finally, the concepts are open-ended: they don't dictate how to accomplish the goals, only what the goals are. For our children's benefit, researchers need to band together around a shared framework, with curricular reform itself treated as a divergent thinking problem: the aim not to find one right answer, but to speak with one voice about values and goals, and present as many viable solutions as possible.

Keywords Creativity · Education · Divergent thinking · Equity

1.1 Introduction

A child entering kindergarten in 2022 won't graduate from a 4-year degree program until 2039. What will the labor market look like then? While there is vigorous debate about whether the effects will be harmful or benevolent, few dispute that industrialized countries are advancing towards ever increasing automation: tasks that can be delegated to a robot or computer will be (McAfee and Brynjolfsson 2016).

A. Brandt (✉)
Shepherd School of Music, Rice University, Houston, TX, USA
e-mail: abrandt@rice.edu

D. Henriksen, P. Mishra (eds.), *Creative Provocations: Speculations on the Future of Creativity, Technology & Learning*, Creativity Theory and Action in Education 7,
https://doi.org/10.1007/978-3-031-14549-0_1

Especially for repetitive or dangerous tasks, human performance is no match for a tireless, relentless, fast-moving machine. By the time today's youngsters start looking for work, many jobs will be manned by robots. Some studies estimate that close to 50% of the United States labor force will be supplanted (Frey and Osborne 2017).

Are we preparing our children for that economy? Current public school education has invested heavily in achieving measurable outcomes in literacy, rote learning, and calculation. But there is a critical piece missing. As Corazza (2016) writes:

> There should be no doubt that the centrality of humans in the future will depend fundamentally on their… cognitive flexibility, capacity to take decisions based on incomplete information, intuition, problem solving ability, artistic and aesthetic sense: in a word, on their creativity (p. 259).

Many see a creative economy as humanity's best bet for future employment (Frey and Osborne 2017). Yet fostering creativity in our young seems less and less of a priority. The scale of the problem is daunting: there are currently over 50 million children in American public schools. The majority suffer from lack of opportunity, with "exposure to innovation during childhood" having "significant causal effects on children's propensities to invent" (Bell et al. 2019, p. 647). Obstacles to improvement are considerable: overflowing classrooms; limited budgets and time; restrictive curricula (Makel 2009). As a result, efforts to bring more creativity into classrooms largely depend on the initiative of individual teachers, often receiving little institutional support. For those who value creativity in the classroom, there are lots of reasons to be discouraged.

Yet in spite of the obstacles, educating our children for the world they will live in should be an endeavor that is too big to fail. Here, cognitive science has an important role to play. As Makel (2009) writes: "The greatest impact creativity researchers could have on society (and where I hope the field is headed) is helping schools foster creative development in our youth" (p. 38).

Similarly, Corazza (2016) asserts that:

> all creativity researchers share a very important social responsibility:… the education of both younger and older generations in the subject of creative thinking and its practical application in all domains of knowledge…Given this sense of urgency and necessity for action…it is of utmost importance to resolve any fragmentation (p. 259).

But how to mobilize a determined effort? There are competing models of the creative process. Open questions abound: the degree to which creativity is domain-general or domain-specific, whether there is "far transfer" between disciplines, whether creativity can be taught, even whether children can be considered creative. Makel and Corazza have issued a call to arms—but can a field whose subject matter celebrates novel thinking and alternative viewpoints and has so many unresolved issues coalesce around unified guidance?

The answer may be hiding in plain sight. From the earliest days of creativity research, *divergent thinking*—the proliferation of multiple options—has been a common denominator in models of the creative process. Torrance (1970), Guilford (1984), Urban (1997), Cropley (2001), and Vincent, Decker & Mumford (2002) are among the many who have argued for its relevance. Whether it is referred to as ideation, brainstorming, or blind variation and selective retention, *divergent thinking* captures the diversification of options that propels human ingenuity. Reviewing the literature, Runco and Acar (2012) conclude: "divergent thinking is not synonymous with creativity but...there is great value in the concept" (p. 9). Historically, a multitude of creativity training programs in both education and industry have shared divergent thinking as a common foundation (Scott et al. 2004). It is the area of greatest consensus.

Building a creativity curriculum around divergent thinking has a host of advantages. First, its terms are straight-forward and easy to define. Second, it has been extensively studied: there are literally decades of research investigating it. It is solidly mainstream, making it easier to persuade school boards and state legislatures. Third, it is trans-disciplinary, as applicable to STEM fields as to the arts. Fourth, abundant exemplars exist (Scott et al. 2004). Fifth, there are objective ways to measure it. Finally, and most crucially, the concepts are open-ended: they don't dictate how to accomplish the goals, only what the goals are.

Still, a tipping point in which creativity becomes a widespread instructional mandate has yet to be reached. If anything, it is slipping backwards. If we fail to meet this moment, we may undermine generations of children. How can what we've learned about the brain and education help to achieve the necessary progress?

1.2 What Is Creativity?

In advocating for creativity in schools, one big stumbling is that the word itself can mean different things to different people. Are we talking primarily about the arts? Is something only creative if it is judged so by experts? Are failed ideas creative?

Creativity can be understood as the production of novel output whose viability and lifespan will depend on perceptions of its value, usefulness, intelligibility, and originality. The weightings of these often subjective judgments can vary from work to work, person to person, field to field, and over time. Da Vinci's *Mona Lisa* is a case in point: in the nineteenth century, it was considered one of the Louvre's lesser holdings. When the painting was stolen off the wall of the museum in 1911, Parisians "lost a masterpiece they didn't really know they had" (Sassoon 2001, p. 12). Impelled in part by the publicity surrounding its theft and safe return, the *Mona Lisa* gradually became the most prized and talked-about painting of all time (Sassoon 2001).

Kaufman and Beghetto's 4C model of creativity (2009) compares the reach and impact of creative work. Mini-c and little-c creativity involve output that is personally meaningful but has limited exposure. Pro-c refers to creative work that receives professional approval; and Big-C creativity refers to achievements that stand the test of time. The 4C model makes the case that creativity is not just represented by the peaks: it is the full mountain range, from the foothills to the summits, covering a vast landscape of human behavior.

Novelty too comes in varying degrees. "Sustaining" creativity refers to output that perpetuates existing standards. For instance, in many artistic traditions, an individual performer adds their own flourishes to inherited melodies, stories, visual art, or dance: the goal is not to improv, upend or replace traditional practices but rather to keep them viable. A sequel is another example of "sustaining" creativity: it takes inventiveness to keep a story going; but that continuation lives comfortably within an existing framework, such as Sherlock Holmes' London or the "Star Trek" universe. Often, chronology is less crucial: for instance, you can read Sherlock Holmes stories out of order.

"Incremental" creativity refers to gradual improvements or evolution: making a car safer or more fuel efficient, building a faster CPU, creating a new spin on crime fiction. Links to the past are palpable: you know you're getting a next-gen product or version 2.0. Here the arrow of time matters more: you're less likely to trade backwards for a more dangerous car or slower computer; a forensic detective can't be backdated to the nineteenth century.

Finally, "disruptive" creativity refers to work that is radical and norm breaking. At its most extreme, there is a clear dividing line of "before" and "after:" before and after Einstein's theory of relativity, Picasso's *Les Demoiselles d'Avignon,* and the world wide web.

Humans generally prefer gradual transitions to sudden overhauls. That's why software developers populate our devices with digital bookshelves, trash cans, and camera shutters; they keep us tethered to the familiar (Brandt and Eagleman 2017). Indeed, as Henrich (2020) argues, the premium Western industrialized societies place on disruptive creativity is largely an outlier in human history. For many cultures past and present, "novelty" is a refresh, not a restart: they're not looking for the next Silicon Valley. Furthermore, disruptive innovations often take flight after a long runway: for example, the electric car has been under development since the birth of the automobile (Høyer 2008). There's no doubt innovation has accelerated in the digital age; but sudden breakthroughs are the exception, not the rule. Most creativity tills the soil of the familiar, replenishing and reinvigorating it.

What does this all mean for educators? First, human creativity is far-ranging: there's almost nothing in our interests and behavior we don't apply it to; it is a hallmark of our species (Brandt and Eagleman 2017). Second, active imaginations generate creative output that ranges from the unviable and short-lived to the cultural treasured and long-lasting. Some ideas are left unrecorded, like a night of theater improve; others leave their mark on history. Some are readily dismissed, even by the person who thought them up; others continue to stir hearts and minds centuries after they were conceived. Vagaries of time and place play an outsize role in how work is received: something treated as irrelevant in one context may be viewed as groundbreaking in another; judgment can be provisional and consensus hard to reach (Brandt 2021). As a result, we need to prepare students to be responsive to a whole range of social and cultural situations. If we only train them to be innovators, they may stumble in more constrained situations; on the other hand, if we only train them to be incrementalists, they may fall short in challenges that demand more radical re-thinking.

All of this points to divergent thinking as a fundamental educational goal: it is through idea diversification that students gain the flexibility and resilience to respond to a complex, ever-changing world.

1.3 Creative Thinking in the Brain

Brains operate in a perpetual tug-of-war between automation and innovation: either streamlining for efficiency and accuracy, or networking for greater flexibility. When we need a fixed outcome, our neural processing is consolidated into dedicated networks, enabling us to reliably do everything from eating and breathing to spelling and math tables (Eagleman 2011). In contrast, when engaged in creative tasks, functional connectivity between brain regions is enhanced and neural networks become less stable (Beaty et al. 2018). Crosstalk between neural regions is how the brain does R&D: the brain accepts a loss of speed and accuracy for the opportunity to surprise.

We now know that our brains are in a perpetual state of remodeling throughout our lifetimes (Bagarinao et al. 2019). Furthermore, neural networks are in a "chronic and unforgiving competition" (Eagleman 2020, p. 45)—because the greater the amount of territory devoted to a task, the better the performance. As far as the brain is concerned, *we are what we do*: for instance, the brain regions involved in spatial navigation are enlarged in London taxi drivers; and the same is true of the motor cortices of musicians, who need fine motor control to play their instruments. What about creativity? Longitudinal studies are rare: however, initial indications are that the same principle holds. For instance, Sun et al. (2016) imaged participants before and after a 20-session course in divergent thinking. The researchers found that improved originality and fluency were correlated with changes in functional connectivity and gray matter volume.

Research has also shown that creative thought involves an intricate dance between focused and unfocussed attention, with a large share of the labor occurring behind the scenes in the unconscious. Salience, goal-directedness, and mind wandering all collaborate (Goldberg 2018). Finally, the serial order effect is well established: responses to a prompt tend to get more unusual as time passes (Beaty and Silvia 2012).

What does all this tell us about creative cognition? Just as society divides up its labor between routine and more bespoke tasks, so too do our brains. Divergent thinking results when neural networks energetically intertwine. It is too reductive to say that divergent thinking alone accounts for the creative process: the options it produces may not be unusual; other mechanisms are involved in deciding among them and making a conclusive choice; and there are times when the brain may arrive directly at a novel solution without considering alternatives. But while divergent thinking isn't the entire machinery of creativity, it is its engine. It is hard to imagine how streamlined behavior alone could regularly generate novelty, since it is designed to do the opposite. A brain that proliferates options has a much higher chance of producing an original result. Furthermore, the serial order effect and the role of undirected thinking suggest that cultivating the creative process should generally not be a sprint: when it comes to surprising ideas, it's important to give the brain *time*.

1.4 Giftedness in Children

Do our genes play a determinant role in creative ability? After all, if creativity is largely innate, can education really improve it? Comparing identical twins is an apt way to address this question: if heredity were a dominant factor, twins would perform similarly on creative tasks, even if raised apart. Yet multiple studies (Richmond 1966; Kandler et al. 2016) have found only a modest correlation: when the environment diverges, so too do twins' development. Similarly, creativity does not seem to consistently run in families: studies "have generally found little evidence for the familial aggregation of creative talent" (Waller et al 1993, p. 235).

As a result, we may be making a crucial mistake when we treat giftedness as reflected primarily in whether or not something comes easily to a child. As the "10,000 rule" suggests, professional attainment is often dependent upon the amount of attention and effort someone is willing to devote to a task (Ericsson et al 2007). For instance, Klissouras et al. (2001) found that twins rarely reach the same level of success—effort matters. Other evidence increasingly points to the environment and training playing decisive roles. For instance, Heinrich (2020) reports that 120 genes that "can potentially influence educational attainment have dropped in frequency during the 20th century"; yet "this genetically induced pushed *against* schooling. . .was rolled over by cultural evolution," (p. 482) which has significantly *raised* schooling outcomes. Natural selection can be overruled by culture.

This raises caution flags about creativity testing, especially for the very young: children have only limited experience, and different degrees of exposure, encouragement, and opportunity. Advanced programs for more precocious youth are certainly reasonable, but creative learning should be mandated for everyone. As a society, we have been too preoccupied with giftedness and not enough with equity.

1.5 Proximal Development

In the early twentieth century, the psychologist Jean Piaget developed a framework for the maturing child that viewed cognitive development as a "preordained unfolding driven by hard-wired intrinsic biological rules" (Goldberg 2018, p. 147). At the same time, Russian psychologist Lev Vygotsky proposed a contrasting model, arguing that children cannot reach their full potential without help (Wertsch 1984). For any skill, Vygotsky defined the *zone of proximal development* as the difference between what the child could accomplish on their own and "the additional expansion made possible through instruction" (Goldberg 2018, p. 147).

Vygotsky' views were initially overshadowed by Piaget's, but they have gradually become more mainstream. Indeed, his model is not radical: whenever parents hire an SAT tutor or send their children to band camp, it's under the presumption that expert guidance will improve the outcomes. Yet while we readily accept proximal development in everything from learning to drive to mastering a foreign language, there is considerable skepticism about whether it is possible to train creative thinkers. For many, creativity is a gift, and imagination can't be taught.

Moreover, we've come to view school as the place where creativity goes to die. In a well-known study, 1600 school children took the same divergent thinking test given to NASA scientists (Land and Jarman 1992). Ninety-eight percent of the 3–5-year-olds scored in the genius category. When the same group took the test 5 years later, "only thirty-two percent scored that high." Five years after that, "it was down to ten percent" (p. 153). More bad news arrived in 2011, when Kyung Hee Kim's study documenting a steady decline in creativity scores over several generations of US school children made the cover of *Newsweek* magazine.

In order to turn the tide and persuade legislatures and school boards that classroom creativity is a worthy investment, pointing to decline is not enough; evidence of positive outcomes is vital. Examples of creativity *going up* through schooling exist, but they are less well-known. For instance, a series of studies in the 1970s offered encouraging results about workbooks designed by Myers and Torrance to stimulate divergent thinking. Nevertheless, in a literature review, Mansfield et al. (1978) find the scarcity of research on the workbooks "surprising" (p. 529). In another meta-study, Scott et al. (2004) reviewed over 70 programs for children and adults and reached "an unambiguous conclusion. Creativity training works" (p. 382). More recently, a study by Sun et al. (2020) found that Chinese high school students who took a course designed to enhance divergent thinking significantly outperformed the control group in measures of creative thinking. Still, more evidence is needed, as well as louder broadcasts when positive outcomes occur. Demonstrating the value and efficacy of creativity curricula needs to become a concerted effort—another area where uniting around a common goal is necessary to make an impact. If public school administrators aren't convinced that creativity training works, why should they try?

1.6 Does Creativity Belong in School?

Adding creativity to the curriculum risks taxing already overworked instructors. As Makel (2009) writes, "Teachers, through high stakes testing and accountability, feel increasingly forced to sprint through as much material as quickly as possible. . .(That is) not conducive for creative experiences because the creative environment is often not viewed as efficient" (p. 40). On top of that, creativity demands a level of personal attention that is challenging in overcrowded classrooms. Nor is school-based creativity necessarily a good fit for students: some psychologists have argued that the classroom may not be the proper forum for an activity that requires self-reliance and experimentation free from prying eyes (Runco et al. 2017). Addressing these concerns requires strategic workarounds: for instance, it may often be preferable to assign creative tasks as homework, where the school bell isn't an issue and there is more privacy. Students can be self-directed in choosing their projects, work in groups to reduce the teacher's load, and be mentored in how to offer constructive feedback to each other.

The arts are typically the standard-bearers for creativity in the curriculum: studies are still scant, but those that do exist have found that historically underserved populations benefit the most from high levels of arts engagement and learning (Catterall et al. 2012; Bowen and Kisida 2019). The rewards are tangible: fewer disciplinary problems; better attendance; higher college enrollment. The paucity of extant research is due in no small measure to the shortage of public school arts programs to study—a reminder of how much this is an issue of equity.

1.7 Designing a Creative Curriculum

Makel (2009) describes the "disparity between valuing creative performance in adults and not fostering creativity in children as the *creativity gap*" (p. 38). There is a paradox involved in trying to close that gap: on the hand, it demands a scalable program, capable of addressing the needs of millions of children; on the other hand, how can there be a one-size-fits-all curriculum that allows for fresh thinking, non-conformity, and an attention to individual needs and strengths?

The answer lies in designing a rigorous framework that allows for significant customization. In applying creativity research to education, *divergent thinking* asserts itself as a critical component: it is well-articulated yet open-ended. It applies to any discipline; and by exploring questions without an answer key, it is a worthy complement to standardized testing driven curricula.

1.7.1 Divergent Thinking 2.0

Certain fundamentals should underlie a creativity curriculum. First, it should not be a "one-off;" instead, it needs to be an extended program carried through all grade levels. Second, it should not be limited to the arts; rather, it should be transdisciplinary (Sawyer 2010). Third, the more varied the creative projects, the more flexible thinkers they are likely to produce. Finally, divergent thinking should be encouraged as much as possible.

How to do that? Divergent thinking is typically described by four main attributes: fluency (the number of responses), elaboration (how well worked out they are), flexibility (how diverse), and originality (how rare or unusual) (Guilford 1968; Torrance 1995). All of these can be taught and practiced.

For instance, consider the design of a building—a task at the cross-roads of engineering and art. Fluency and elaboration involve drafting multiple models—something routine in any architectural firm. To motivate flexibility and originality, students can be prompted to develop a range of plans: some that blend in with neighboring buildings, others that include modest updates, and still others that boldly clash with their surroundings. Every aspect of this process can be mentored.

There are a host of other ways to incorporate divergent thinking into the classroom. For instance, real world problems without a solution can engage students on issues of immediate relevance. For instance, Huang et al. (2010) used computer gaming to encourage students to proliferate options in response to a chronic problem afflicting Taiwan: the hazardous debris created by landslides. The researchers found that their playful approach successfully impelled students to work out multiple solutions.

Any subject can be enriched by free-wheeling questions. Gallavan and Kottler (2012) integrated divergent thinking into the social studies curriculum, encouraging students to consider history from multiple angles, and asking them to propose economic and political innovations. Kwon et al. (2006) likewise designed open-ended problems for middle school math: in a controlled study, they found that the treatment group showed particularly impressive results in originality.

One way to stimulate divergent thinking is to alter or loosen constraints. Science fiction prototyping (Johnson 2011) and alternative worlds and histories are particularly potent ways to encourage originality and flexibility: they require students to navigate hypothetical scenarios. What if the electoral college were abolished? What if self-driving automobiles were capable of real-time coordination? What if you could add more one sense—what would it be and why (Eagleman 2020)?

Because arts classes typically don't have to "teach to the test," they have enhanced opportunities to stimulate divergent thinking. It all boils down to a simple lesson: *think in multiples*. Yet many art classes do not explicitly prioritize this. Often this is because of resource scarcity: not enough materials or time. While that's certainly understandable, it is nevertheless a missed opportunity. Furthermore, divergent thinking isn't fully leveraged if solutions are too similar. Instead, students should be prompted to explore a spectrum of solutions from the familiar to the far-out (Brandt and Eagleman 2017). Once again, the arts are an ideal laboratory, because they can be more easily unbound from practical constraints.

Take a theme and variations, Western classical music's most overt expression of *divergent thinking*: a melody, often familiar, is revisited over and over but always in a new way. The number of variations is an index of the composer's imagination: the more, the better. Crucially, as the variations progress, they tend to get further and further removed from the source (Brandt 2019).

Creating variations is readily adaptable to other media. For instance, in an undergraduate creativity course I teach at Rice University, I ask students to write a theme and variations on a poetry reading. The requirements are that they write at least four variations; the variations should be as distinct from each other as possible;

and the source should become less and less recognizable. One student based her set on Langston Hughes' "Mother to Son":

> Well, son, I'll tell you:
> Life for me ain't been no crystal stair ... (Rampersad and Roessel 1995, p. 30)

Her final variation was a sculpture of a staircase on which lines from the poem were inscribed.

Another based hers on a poem celebrating the search for alien life: her final variation was a transcription of the poem into chemical symbols, as is done on Voyager's golden record. In order to encourage risk-taking, students are graded solely on objective measures spelled out in advance: did they create at least four variations? Are they contrasting? Do they go different distances from the source? That way, students can try out wacky ideas without being worried about being penalized. The farther the students' sets get from their sources, the farther they tend to diverge from each other: "far" variations have ranged from audio tracks to videos, visual art, games, puzzles, even recipes—illustrating that retreating from the source encourages greater originality.

The overall goal should be a curriculum that is varied, unpredictable, increasingly challenging for higher grade levels, and personalized to the needs, values, and traditions of the students. Rather than trying to identify a single approach, aggregating a host of well-vetted ones—like a well-managed stock portfolio—is a better strategy: it will increase the number of stakeholders and minimize competition and conflict; be more adaptive to different learning styles; and, in the end, better represent the concepts it is trying to instill.

For all of its value, cultivating divergent thinking alone isn't enough: encouragement of self-expression, risk-taking, and tolerating ambiguity are all crucial (Urban 2003). As Cropley (2001) writes, "the optimum conditions for creativity exist when all of the dimensions (e.g. the properties of the person, the situation, the task, and the solution) are favorable" (p. 146). The more an entire school is involved in creative instruction, the more likely these other conditions will be met.

Implementing such a broad overhaul will be a heavy lift. The alternative is the more ad hoc approach that we have now. The risks there are that it has been hard to make creativity a priority, accountability is difficult, teacher turnover can be destabilizing, and results are almost impossible to compare.

As with any ambitious undertaking, the devil is in the details. Instructors need to be trained, and time needs to be set aside in the busy school day. There's always the risk that something designed to promote flights of fancy may be misapplied or become calcified. But the potential rewards outweigh the obstacles and drawbacks: equity for young people of all backgrounds; classrooms as incubators for tomorrow's flexible thinkers.

As we learn more about the creative brain and different teaching modules are field tested, it will be imperative to revisit curricular guidelines. But after three quarters of a century of research, it's unlikely that divergent thinking will be unseated as a core component. It will take more than these concepts alone to improve student outcomes; but it is a durable platform on which to begin to do so.

1.8 How Technology Can Help

Since the dawn of civilization, technology has given an assist to human creativity: the knife for the carver, the Bunsen burner for the chemist, the telescope for the astronomer. But nothing comes close to the digital world in its impact on creative education. First, the power and availability of computers have democratized creativity: film editing used to require a roomful of high-priced equipment; now it can be done on a laptop. Second, computers are multi-use: the same device can sequence the genetic code and run music notation software, making it an ideal tool for cross-disciplinary learning. Third, digital memory is increasingly affordable and bountiful, making it possible to preserve all drafts. That's important, because the greatest impediment to divergent thinking is having to destroy an earlier version to try out alternatives.

The computer game Minecraft is a good example of a digital platform that has found multiple uses in education. The game offers a sandbox in which players construct virtual worlds using blocks and other shapes. In the game's creative mode, players have limitless access to materials (Bos et al. 2014). There is no winning and losing, just free-form play; and, as with Legos, it is easy to tear down and redesign. Thanks to its simple premise and open-endedness, it has been used to teach math, biology, ecology, physics, and English literature (Short and Short 2012; Bos et al. 2014; Cipollone et al. 2014).

All creative tools involve trade-offs and limitations that condition their output: for instance, for all of its strengths, Western music notation cannot capture the complex pitch and rhythmic inflections of an Indian raga; in those respects, traditional Western scores are simpler. The more students rely on technology for creative work, the more essential it is that these tools are not too prescriptive: otherwise, the "guard-rails" of the software may overly limit young imaginations.

Consider Hyperscore, developed by composer Tod Machover and colleagues at MIT's Media Lab (Farbood et al. 2004). It is a graphic interface for composing that doesn't require knowledge of music notation. Instead of the traditional five-line staves, it offers a painter's palette of dots and lines for pitches, and colors for timbre. Melodies are shaped by brushstrokes, and can be layered at will. Volume is controlled by shrinking or expanding lines and shapes. A "harmony engine" adds traditional tonal harmonies, but it can also be turned off. Finally, there is real time playback. It is possible to transcribe traditionally notated music—a Beethoven Symphony is one of the demos—but it's also possible to go well beyond that. There are constraints: for instance, the melody lines come in eight colors; polyphony beyond that becomes harder to follow. Nevertheless, the program never answers the question "What should music sound like?" Instead, it leaves it to each student to discover their own answers. Platforms such as Hyperscore that are easy to master and make it convenient to generate, test, and save alternatives have the potential to be breakthrough creative learning tools.

1.9 Assessment

As Sawyer (2010) writes, "In today's climate of accountability, what is not assessed does not count" (p. 185). Here again divergent thinking offers distinct advantages: decades of research have been rooted in evaluating it; well-established methodologies offer guidelines for instructors. There are objective measures: a student who only creates one or two variations needs to work on their fluency; one whose answers are too similar or routine need to stretch their flexibility and originality. For a comprehensive measure, a full battery of assessments is recommended, including domain-specific testing, juried appraisals (Amabile et al. 2019), and classroom observation. That may be labor intensive, but the effort put into it will also raise the profile of creative learning at the school.

Moreover, incorporating creativity into course-based assessments may help to address long-standing issues of bias and equity. The national STEM program "Project Lead the Way" (PLTW) added measures of "transportable skills" such as divergent thinking to their standardized tests administered to over 335,000 students. They found that including the transportable skills help to mitigate the historic achievement gap for marginalized students, especially with respect to gender (Gough and Williams 2020). This further incentivized PLTW to teach these skills. Thus, a virtuous feedback loop was created between curriculum and assessment: the more creativity was reflected in both, the higher a priority it became.

1.10 The Importance of Advocacy

A recent study (Lopez Turley 2016) found that, 50 years after a landmark report laying out the inequities faced by minority students, efforts to close those gaps have largely failed. Why? More than any other factor, the study places the blame on the disconnect between researchers and policymakers. Academic faculty are generally not rewarded for engaging with the public; meanwhile, school districts are hesitant to share privileged data for fear of being maligned or misrepresented. As a result, the two communities rarely intersect.

It is not a stretch to imagine that similar problems affect creativity in education. A meta-study by Forgeard and Kaufman (2016) found that the majority of creativity studies they reviewed "offered little or no discussion explaining why readers should care" (p. 255). The authors reason that a primary reason for this may be that researchers are primarily speaking to each other. As important as basic research is, the authors argue that it is important to reach outwards, articulating creativity's benefits to the public and explaining why and when it should supersede other priorities, especially given limited budgets and hours in the school day. If researchers don't want schools to be creativity deserts, we need to engage with policymakers more often.

Change won't happen overnight, so it is vital to develop 5 and 10-year plans, highlighting what can be accomplished right away, and what will take longer to bring to fruition. That takes putting many heads together. Like any field, creativity research is fragmented: researchers from different institutions have few opportunities to collaborate or coordinate. On top of that, creative inquiry is multi-disciplinary, but disciplines too often remain siloed. Academic goals and incentives need to be revisited: there needs to be more interdisciplinary opportunities and greater rewards for the collective effort that real world engagement requires. We know enough now to change children's lives; the time has come to put that knowledge to good use on a bigger scale. This itself should be treated as a divergent thinking problem: the aim not to find one right answer, but to speak with one voice about values and goals, and present as many viable solutions as possible.

1.11 Conclusion

Thanks to a preponderance of evidence, scientists have coalesced around a rigorous understanding of our warming climate, and have become activists in raising the alarm, warning about imminent consequences, and working assiduously to find solutions. It is time to make the creative climate in our classrooms a higher priority. While it is never appropriate to have a party line in science, it is not hard to imagine cognitive scientists sounding similar warnings about how we are educating our children. After many decades of research, the majority would agree: all children have creative potential; creativity depends at least as much on exposure and opportunity as on hereditary factors; it can be improved through instruction and practice; and we are not doing enough. Inequities in access to innovation have increasingly become public health and civil rights issues—a shameful neglect of millions of young minds. As with environmental threats, the threats to livelihood and well-being posed by automation are looming, and waiting too long to address them may cause irreversible harm.

No one study or article on its own is likely to move the needle, no single voice enough to be a wake-up call. As with climate change, what is needed now is coalition-building, coordinated research, and broad-based advocacy based on shared data and guidelines. After many decades of research, there is enough common ground to create robust curricular guidelines. Divergent thinking may be one of the oldest concepts in creativity research, but it has withstood the test of time, offering the opportunity for varied solutions built on shared scaffolding. It is a theme capable of unending variations.

Generating multiple solutions is not the whole picture of creativity: *convergent thinking* plays a complementary and essential role (Dietrich 2015). But students get plenty of practice zeroing in on a correct answer. What's too often missing for them is even a hint of divergent thinking.

Nor is future employment the sole reason for cultivating creativity. Exercising our creative faculties is personally fulfilling and meaningful, promotes empathy and curiosity, and is both challenging and fun. A student prepared for the future economy will also have tools to express themselves, sustain long-term relationships, and raise their children.

Neuroscience and psychology are converging on a view of the brain that is largely shaped by experience and training (Eagleman 2020). The adage that teachers are "molding young minds" is turning out to be literally true: the more time and effort we spend on a task, the more brain territory is devoted to it; that, in turn, is reflected in measurable improvements and accomplishments. Given how quickly the digital economy can remake itself and how much it will intrude on human work, future adults will need to be inventive, mentally agile, resilient, and adaptable. We need to ensure that young people of all backgrounds grow into those kinds of adults. If we want to prepare our kindergarteners for the world that awaits them, the time is now to answer that call.

References

Amabile, T. M., Collins, M. A., Conti, R., Phillips, E., Picariello, M., ... Whitney, D. (2019). A Consensual Technique for Creativity Assessment. In *Creativity in Context* (pp. 41–79). Routledge. https://doi.org/10.4324/9780429501234-3

Bagarinao, E., Watanabe, H., Maesawa, S. *et al.* (2019). Reorganization of brain networks and its association with general cognitive performance over the adult lifespan. *Sci Rep* **9,** 11352. https://doi.org/10.1038/s41598-019-47922-x

Beaty, R. E., & Silvia, P. J. (2012). Why Do Ideas Get More Creative Across Time? An Executive Interpretation of the Serial Order Effect in Divergent Thinking Tasks. *Psychology of Aesthetics, Creativity, and the Arts*, 6(4): 309–319. https://doi.org/10.1037/a0029171

Beaty, R. E., Kenett, Y. N., Christensen, A. P., Rosenberg, M. D., Benedek, M., Chen, Q., ... Silvia, P. J. (2018). Robust prediction of individual creative ability from brain functional connectivity. *Proceedings of the National Academy of Sciences of the United States of America, 115*(5), 1087–1092. https://doi.org/10.1073/pnas.1713532115

Bell, A., Chetty, R., Jaravel, X., Petkova, N., & Van Reenen, J. (2019). Who becomes an inventor in America? The importance of exposure to innovation. *Quarterly Journal of Economics, 134*(2), 647–713. https://doi.org/10.1093/qje/qjy028

Bos, B., Wilder, L., Cook, M., & O'Donnell, R. (2014). Learning mathematics through Minecraft. *Teaching Children Mathematics, 21*(1), 56–59.

Bowen, D.H. & Kisida, B. (2019). Investigating Causal Effects of Arts Education Experiences: Experimental Evidence from Houston's Arts Access Initiative (Volume 7, Issueno 4).Rice University's Kinder Institute for Urban Research.

Brandt, A. & Eagleman, D. (2017). *The Runaway Species: How Human Creativity Remakes the World.* New York: Catapult.

Brandt, A. (2019). "Theme and variations as a window into the creative mind." *Mobile Brain-Body Imaging and the Neuroscience of Art, Innovation and Creativity.* New York: Springer.

Brandt, A. (2021). Defining Creativity: A View from the Arts. *Creativity Research Journal.* https://doi.org/10.1080/10400419.2020.1855905

Catterall, J. S., Dumais, S. A., & Hampden-Thompson, G. (2012). The arts and achievement in at-risk youth: Findings from four longitudinal studies. Washington, DC: National Endowment for the Arts.

Cipollone, M., Schifter, C. C., & Moffat, R. A. (2014). Minecraft as a creative tool: A case study. *International Journal of Game-Based Learning, 4*(2), 1–14. https://doi.org/10.4018/ijgbl.2014040101

Corazza, G.E. (2016). Potential Originality and Effectiveness: The Dynamic Definition of Creativity. *Creativity Research Journal, 28*(3), 258–267. https://doi.org/10.1080/10400419.2016.1195627

Cropley, A.J. (2001). *Creativity in Education & Learning: A Guide for Teachers and Educators.* London: Kogan Page Limited.

Dietrich, A. (2015). *How Creativity Happens in the Brain.* New York: Palgrave Macmillan.

Eagleman, D. (2011). Incognito: The Secret Lives of the Brain. New York, NY: Pantheon Books.

Eagleman, D. (2020). Livewired: The Inside Story of the Ever-Changing Brain. New York, NY: Pantheon Books.

Ericsson, K.A., Roring, R.W. & Nandagopal, K. (2007). Giftedness and evidence for reproducibly superior performance: an account based on the expert performance framework. *High Ability Studies, 18 (1)*: 3–56.

Farbood, M.M., Pasztor, E. and Jennings, K. (2004). "Hyperscore: a graphical sketchpad for novice composers." *IEEE Computer Graphics and Applications*, vol. 24, no. 1, pp. 50–54. https://doi.org/10.1109/MCG.2004.1255809.

Forgeard, M. J., & Kaufman, J. C. (2016). Who cares about imagination, creativity, and innovation, and why? A review. *Psychology of Aesthetics, Creativity, and the Arts, 10*(3), 250.

Frey, C. B., & Osborne, M. A. (2017). The future of employment: How susceptible are jobs to computerisation? *Technological Forecasting and Social Change, 114*, 254–280. https://doi.org/10.1016/j.techfore.2016.08.019

Gallavan, N.P. & Kottler, E. (2012). Advancing Social Studies Learning for the 21st Century with Divergent Thinking. *The Social Studies,* 103:4, 165–170. https://doi.org/10.1080/00377996.2011.605641

Goldberg, E. (2018). *Creativity: The Human Brain in the Age of Innovation.* New York: Oxford University Press.

Gough, M., & Williams, A. (2020, April 30). Measuring creative thinking via holistic standardized assessments: from operationalization to impact on equity. https://doi.org/10.35542/osf.io/qp2t7

Guilford, J.P. (1968). *Creativity, intelligence, and their educational implications.* San Diego, CA: EDITS/Knapp.

Guilford, J. P. (1984). Varieties of divergent production. *The Journal of Creative Behavior, 18*(1), 1–10. https://doi.org/10.1002/j.2162-6057.1984.tb00984.x

Henrich, J. H. (2020). *The weirdest people in the world: How the west became psychologically peculiar and particularly prosperous.* New York, NY: Farrar, Straus and Giroux.

Høyer, K.G. (2008). The history of alternative fuels in transportation: The case of electric and hybrid cars, *Utilities Policy, 16 (2),* 63-71. /https://doi.org/10.1016/j.jup.2007.11.001.

Huang, C. C., Yeh, T. K., Li, T. Y., & Chang, C. Y. (2010). The idea storming cube: Evaluating the effects of using game and computer agent to support divergent thinking. *Educational Technology and Society, 13*(4), 180–191.

Johnson, B.D. (2011) Science Fiction Prototyping: Designing the Future with Science Fiction. San Rafael, California: Morgan & Claypool.

Kandler, C., Riemann, R., Angleitner, A., Spinath, F. M., Borkenau, P., & Penke, L. (2016). The nature of creativity: The roles of genetic factors, personality traits, cognitive abilities, and environmental sources. *Journal of Personality and Social Psychology, 111*(2), 230–249. https://doi.org/10.1037/pspp0000087

Kaufman, J.C. & Beghetto, R.A. (2009). "Beyond Big and Little: The Four C Model of Creativity." *Review of General Pyschology,* 13, 1: 1–12.

Kim, K. H. (2011). The Creativity Crisis: The Decrease in Creative Thinking Scores on the Torrance Tests of Creative Thinking. *Creativity Research Journal, 23*(4), 285–295. https://doi.org/10.1080/10400419.2011.627805

Klissouras, V., Casini, B., Di Salvo, V., Faina, M., Marini, C., Pigozzi, F., Pittaluga, M.,Spataro, A., Taddei, F. & Parisi, P. (2001) Genes and Olympic performance: a co-twin study, International Journal of Sports Medicine, 22, 250–255.

Kwon, O. N., Park, J. H., & Park, J. S. (2006). Cultivating divergent thinking in mathematics through an open-ended approach. *Asia Pacific Education Review*, *7*(1), 51-61.

Land, G. & Jarman, B. (1992). *Breakpoint and Beyond: Mastering the Future–Today*. New York: Harper Collins.

López Turley, R. N. (2016, September 1). Connecting research and policy to reduce inequality. *RSF*. Russell Sage Foundation. https://doi.org/10.7758/rsf.2016.2.5.13

Makel, M. C. (2009). Help Us Creativity Researchers, You're Our Only Hope. *Psychology of Aesthetics, Creativity, and the Arts*, *3*(1), 38–42. https://doi.org/10.1037/a0014919

Mansfield, R. S., Busse, T. V., & Krepelka, E. J. (1978). The Effectiveness of Creativity Training. *Review of Educational Research*, *48*(4), 517–536. https://doi.org/10.3102/00346543048004517

McAfee, A., & Brynjolfsson, E. (2016). Human work in the robotic future: Policy for the age of automation. *Foreign Affairs*, *95*(4), 139-150.

Rampersad, A., & Roessel, D. (Eds.). (1995). *The Collected Poems of Langston Hughes*. New York: First Vintage Classics.

Richmond, B. O. (1966). Creativity in Monozygotic and Dyzygotic Twins.

Runco, M. A., & Acar, S. (2012). Divergent thinking as an indicator of creative potential. *Creativity research journal*, *24*(1), 66-75.

Runco, M. A., Acar, S., & Cayirdag, N. (2017). A closer look at the creativity gap and why students are less creative at school than outside of school. *Thinking Skills and Creativity*, *24*, 242–249. https://doi.org/10.1016/j.tsc.2017.04.003

Sassoon, D. (2001). "Mona Lisa": The Best-Known Girl in the Whole Wide World. *History Workshop Journal*, 51: 1-18.

Sawyer, K.S. (2010). Learning for Creativity. In R.A. Beghetto & J.C. Kaufman (Eds.), *Nurturing Creativity in the Classroom*. Cambridge, England: Cambridge University Press.

Scott, G., Leritz, L. E., & Mumford, M. D. (2004). The Effectiveness of Creativity Training: A Quantitative Review. *Creativity Research Journal*, *16*(4), 361–388. https://doi.org/10.1207/s15326934crj1604_1

Short, D., & Short, B. D. (2012). Teaching scientific concepts using a virtual world–Minecraft. *Teaching Science: The Journal of the Australian Science Teachers Association*, *58*(3), 55–58. Retrieved from http://search.ebscohost.com/login.aspx?direct=true&db=ehh&AN=83256656&site=ehost-live

Sun, J., Chen, Q., Zhang, Q., Li, Y., Li, H., Wei, D., ... Qiu, J. (2016). Training your brain to be more creative: brain functional and structural changes induced by divergent thinking training. *Human Brain Mapping*, *37*(10), 3375–3387. https://doi.org/10.1002/hbm.23246

Sun, M., Wang, M., & Wegerif, R. (2020). Effects of divergent thinking training on students' scientific creativity: The impact of individual creative potential and domain knowledge. *Thinking Skills and Creativity*, *37*. https://doi.org/10.1016/j.tsc.2020.100682

Torrance, E.P. (1970). *Encouraging Creativity in the Classroom*. Dubuque, IA: Wm. C. Brown Publishers.

Torrance, E.P. (1995). *Why fly?* Norwood, NJ: Ablex.

Urban, K. K. (1997). Modelling creativity: the convergence of divergence or the art of balancing. In J. Chan, R. Li & J. Spinks (Eds.), *Maximizing Potential: Lengthening and Strengthening Our Stride* (pp. 39–50). Hong Kong: University of Hong Kong Social Sciences Research Center.

Urban, K. K. (2003). *Toward a componential model of creativity.* In D. Ambrose, L. M. Cohen, & A. J. Tannenbaum (Eds.), *Perspectives on creativity. Creative intelligence: Toward theoretic integration* (p. 81–112). Hampton Press.

Vincent, A. S., Decker, B. P., & Mumford, M. D. (2002). Divergent thinking, intelligence, and expertise: A test of alternative models. *Creativity research journal, 14*(2), 163-178.

Waller, N. G., Blacker, D. M., Bouchard, T. J., Lykken, D. T., & Tellegen, A. (1993). Creativity, Heritability, Familiality: Which Word Does Not Belong? *Psychological Inquiry, 4*(3), 235–237. https://doi.org/10.1207/s15327965pli0403_18

Wertsch, J. V. (1984). The zone of proximal development: Some conceptual issues. *New Directions for Child and Adolescent Development, 1984*(23), 7-18.

Anthony Brandt is a Professor of Composition and Theory at Rice University's Shepherd School of Music and Artistic Director of the contemporary music ensemble Musiqa. Dr. Brandt's musical catalogue includes three chamber operas, as well as orchestral, chamber, vocal, theater, dance, and television scores. Dr. Brandt and neuroscientist David Eagleman have co-authored The Runaway Species: How Human Creativity Remakes The World. Dr. Brandt has also contributed chapters to the Oxford Handbooks of Music and the Brain and Music and Language, and published papers in Frontiers, The Creativity Research Journal, Brain Connectivity, and IBRO Neuroscience Reports.

Chapter 2
The Intersection of Human and Artificial Creativity

David H. Cropley, Kelsey E. Medeiros, and Adam Damadzic

Abstract In this chapter we hypothesise that, with regard to the defining activity of creativity (the ability to generate novel and effective outputs), artificial systems are limited to, at best, moderate levels of incremental creativity. In other words, artificial systems have the potential to generate new, and effective, variations of existing ideas, solutions, systems, and artefacts. Furthermore, even with improvements and changes to the technology of AI, this capacity is likely not to transition, eventually, into an autonomous ability for radical creativity, but simply into higher levels of incremental creativity, at a lower cost. While computing power will increase, and algorithms will continue to improve, the limiting factor on artificial creativity (aside from possible data and energy constraints) is not how (the process): rather, it is the rationale (why). No matter how good AI technology becomes, the reason why we are creative (problem definition and solution validation) remains the job of humans. Furthermore, the importance of educating humans for creativity becomes stronger as the jobs of the future coalesce around non-routine, non-algorithmic, non-automatable jobs and tasks.

Keywords Creativity · Artificial intelligence · Humans · Process · Integration

D. H. Cropley (✉)
University of South Australia, Adelaide, Australia
e-mail: david.cropley@unisa.edu.au

K. E. Medeiros
University of Nebraska, Omaha, NE, USA

A. Damadzic
University of Nebraska, Omaha, NE, USA

D. Henriksen, P. Mishra (eds.), *Creative Provocations: Speculations on the Future of Creativity, Technology & Learning*, Creativity Theory and Action in Education 7,
https://doi.org/10.1007/978-3-031-14549-0_2

2.1 Introduction

The rise of Industry 4.0 – the proliferation of cyber-physical systems, artificial intelligence, big data, and automation – has turned attention, once again, to the interaction between humans and robots.[1] Captivating attention in both academic and public spheres, the debate on how humans and robots interact largely centres around the interplay between human and artificial cognition. The human-robot cognition interaction fuels practical inquiries into the formation of high-performing human-robot teams, leveraging robots to enhance human cognition, and the capacity for robots to overtake human cognition. At the heart of these conversations, however, lies a critical question – what does the Future of Work look like? (see, for example, OECD 2017). As robots take on more and more tasks previously performed by humans, where does that leave the human worker?

A common argument that forms the basis of the World Economic Forum's characterisation of twenty-first century *skills* (WEF 2016, 2020) is that robots, and their underpinning technologies (e.g. artificial intelligence), are ideally suited to routine, algorithmic tasks. Thus, robots can be designed to replace humans as bookkeepers, drivers, administrative assistants, laboratory assistants, telemarketers, paralegals, and even bartenders. This leaves humans with the jobs that robots *cannot* do. These uniquely *human* tasks typically revolve around *soft* skills, for example: leadership and emotional intelligence, persuasion, and negotiation, and, critical to the present discussion, creativity and complex problem solving (WEF 2020).

In fact, nearly 60 years ago, the psychologist Jerome Bruner anticipated a future in which "thinking machines" (1962b, p. 6) would take on routine problems (i.e. those that are "well-formed" and "amenable to a unique solution"), leaving creativity, and creative problem solving, as the locus of human dignity and worth (1962a, pp. 2–3). Creativity, in other words, is the bastion of human mental sovereignty over robots. This *all-or-nothing* view is now common, and rather appealing. We have nothing to fear from this perspective, provided we focus our attention on developing our unique capacity for creativity. Indeed, creativity is becoming an important focus of school curricula across the world for precisely this reason (e.g., Patston et al. 2021).

The reality of the Future of Work, however, may be somewhat more complex and nuanced. While psychologists may argue that creativity is a uniquely human ability, we see more and more examples of artificial systems that are claimed to be *creative*. For example, media heralded the win of Google's AlphaGo contestant in its match against the world's top Go player (BBC 2017). Similarly, the production of *1 the Road*, a novel composed by a trained AI system on a road trip, has been praised as a sign of creativity in artificial intelligence (The Atlantic 2018). The impact of the Future of Work may therefore not be as clear-cut as some would suggest. Will humans continue to dominate creativity, complex problem solving and related soft

[1]The term "robot" is used here as a general reference to systems of hardware and/or software that employ artificial intelligence and automation to replicate human actions.

skills, or will robots usurp human sovereignty? Or is there a happy medium in which humans and robots interact productively for a net benefit? In addition, what are the implications of this for education and learning? To answer these questions, it is first necessary to delve into the nature of human creativity and the potential for artificial systems to replicate this capability.

The subject of creativity is governed by two distinct, overlapping, and perhaps competing, disciplines. On the one hand, *Psychology* has spent over 60 years exploring questions about human creativity and: (a) cognition – especially divergent thinking; (b) attitudes and dispositions – for example, personality traits, feelings and motivations; (c) environmental factors – including organisational and classroom climate and culture, and (d) the characteristics of products – the properties that make an artefact or idea *creative*. On the other hand, *Computational Creativity* has, more recently, sought to model and replicate creativity in order to: (a) construct an artificial system capable of creativity; (b) uncover algorithmic descriptions of human creativity, and (c) develop artificial aids to human creativity.

Any discussion of the intersection of human and artificial creativity therefore must draw on, and attempt to reconcile, the concepts, tools, and methods of these disciplines. That there has been relatively little integration of these disciplines – see, for example, the comments relative to computational creativity by Lamb, Brown, and Clarke (2018, p.1) – suggests that this is no simple task. It is perhaps no surprise that Psychology, with its focus on human behaviour, cognition and emotions, has a vested interest in refuting claims that *computers can be creative*, while equally, Computational Creativity has a vested interest in demonstrating that these claims are valid.

As the world grapples with the impact of Industry 4.0 on the Future of Work, what is certain is that the question – *can robots be creative* – must be addressed. The answer to this question has important implications for education and learning, the global economy, and ultimately, human dignity and worth. The following provides guidance on solving these problems by first addressing the human-factored centrality of creativity, creative processes, and integrating AI systems.

2.2 The Psychological Foundations of Human Creativity

The question of whether or not artificial creativity can replicate, in whole or in part, human creativity, first requires that we define the *scope* of this replication. What is it, exactly, that an artificial system must replicate, with respect to human creativity?

Human creativity encompasses four factors, defined by Rhodes (1961) and Barron (1969) as Process (cognition), Person (attitudes and behaviours), Press (climate and culture) and Product (outcome). In simple terms, our ability to generate creative outputs (Product) is a function of how we think (Process), who we are (Person) and where we work (Press).

Two of these factors address uniquely human *limitations* on creativity. Optimism, enthusiasm, tolerance for uncertainty and openness (facets of the Person) are important *internal* factors for creativity because of their capacity to help or hinder humans in their efforts to think creatively (Process). Thus, while an individual might be highly adept at divergent thinking (Process), if they are closed-minded, intolerant of ambiguity and disinterested (Person), then their overall capacity for creativity – their ability to generate a creative Product – is likely to be low. Time, resources, and support (facets of Press), in similar fashion, are important *external* factors for creativity only because they can support or constrain human efforts to be creative. A motivated, open, and flexible individual (Person) strong in divergent thinking (Process) may still fail to generate creative Products if they are embedded in a workplace (Press) that is hostile to new ideas.

In contrast, an artificial system is neither aided nor restrained by personality and environment – indeed, that is surely one of the *attractions* of artificial creativity – and therefore these facets of human creativity do not need to be replicated. Only Process (*how* creative ideas are generated) and Product (the creative ideas themselves) should therefore be in scope for computational systems seeking to replicate human creativity. The centrality of Process and Product is consistent with the thinking of early *computational* pioneers (e.g., Newell et al. 1962), but perhaps at odds with more recent computational notions of creativity (e.g., Wiggins 2006) that seem to focus on replicating the Person: "…*behaviour* [emphasis added] exhibited by natural and artificial systems, which would be deemed creative if exhibited by humans." Artificial systems do not need to mimic the attitudes, behaviours, or actions of the creative human. They must replicate only the cognitive process and the result.

Freed from the uniquely human limitations on creativity (Person and Press), Process and Product define two key constructs that must be understood before human and artificial creativity can be reconciled and perhaps integrated. Product demands that we ask *why* we are creative, while Process requires us to address *how* we are creative.

2.3 *Why* Are We Creative?

It is widely agreed, both in the psychological realm (from Bruner 1962a to Runco and Jaeger 2012) and in the computational realm (from Newell et al. 1962 to Lamb et al. 2018), that to be judged creative, a product[2] must be: (a) novel, and; (b) effective. These two criteria represent the final embodiment of the underlying purpose – the driving force – behind creativity. They explain why we bother with creativity in the first place. Moreover, neither novelty nor effectiveness *alone* are sufficient for a creative product. For instance, solving an overcrowded parking problem by "parking on the moon" is novel, but ineffective given its practicality.

[2]For the sake of simplicity, we use *Product* to mean any idea, *artefact*, *process*, *system* or *service*. Any tangible or intangible output of a creative process.

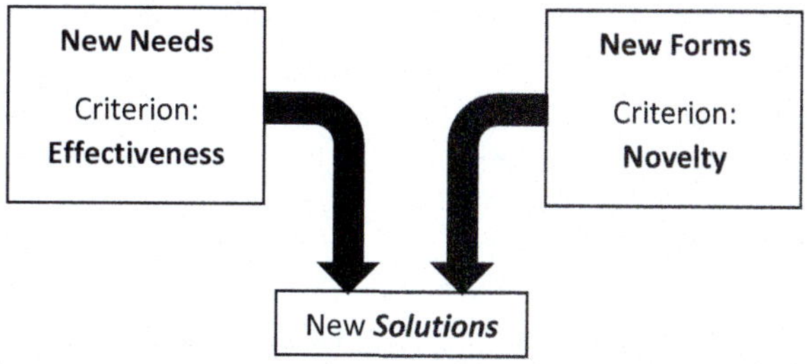

Fig. 2.1 New needs and new forms lead to new solutions

In contrast, developing additional spaces in an existing parking lot is effective but lacks novelty.

In a scientific context, these new forms/products (Fig. 2.1) might be new *artefacts*: for example, an electronic device, or a physical material. In an artistic context, they might be new *compositions*: a picture, a melody, or a story. New needs, in a scientific sense, might be *how to reduce carbon emissions*, while in an artistic sense, they might include *how to capture the beauty of a landscape* (see Fig. 2.1).

Importantly, creativity is not merely the production of novel forms on their own. Nor is it merely the definition of novel problems (needs). Creativity, ultimately, is the production of new artefacts, ideas, compositions, and systems (judged by the criterion *novelty*) that satisfy an identified need or solve a problem (judged by the criterion *effectiveness*). As such, creativity is the generation of a novel and effective *solution* (Fig. 2.1).

The development of this solution may begin with the desire to create a novel form (that must eventually serve a purpose), or it may begin with a desire to solve a new problem (that can only be tackled with a new solution). However, whether we consider the form or the function, the *origin* of creativity – the reason or purpose, if not the first step – is *the human need*. For instance, the Brassiere invented by Mary Jacob in 1914 (Cropley 2020) is anchored in the need for a more comfortable and fashionable undergarment to replace the corset, while the Printing Press invented by Johannes Gutenberg in 1450 (Cropley 2019) is tied to the need share information. Regardless of the pathway followed, there is an underpinning need that is satisfied, and these needs can only be expressed by humans. Lamb et al. (2018) acknowledge as much in their discussion of the lack of *autonomy* (p.7) in computational systems, or "the [in]ability of the machine to decide for itself what to do."

Why, then, are we creative? To satisfy *human* needs. An algorithm could never determine that Mary Jacob found corsets uncomfortable and unfashionable. An AI could never decide that humans had a desire to share and communicate written information more quickly and easily. Even if a computational system can present a report or provide information that signals an emerging gap or need, *human* cognition is still required to decide if that is, indeed, a problem, or if a new form is desirable. Asking *why* we are creative shows us that the first stage of creative problem solving (e.g., Guilford 1959), is *Problem Recognition/Definition*, and is *uniquely* human. As a first step to reconciling, and possibly integrating, human and artificial creativity, we see that the process cannot be entirely separated from human input, or human

Table 2.1 The roles of human and artificial creativity

	Problem definition	Process	Solution validation
Human	✔	✔	✔
AI	✗	???	✗

sovereignty, no matter how sophisticated the artificial means brought to bear. As Guckelsberger et al. (2017) acknowledge, the answer to why an artificial system made a creative decision would always, eventually, return to "because my [human] programmer told me to."

In line with this reasoning, *Solution Validation, defined* as the final stage in the process of creative problem solving by Guilford (1959), represents how well a product does, in fact, resolve the problem or address the need. It is important to note here the distinction between *verification* and *validation*. The former establishes that a solution meets a specification: the car travels at 60 km/h; the kettle heats water to 100 C. The latter, *validation*, is a matter of establishing that the solution *satisfies the human need*: The Brassiere is, in fact, more comfortable than a corset; the Printing Press does, indeed, allow more rapid and widespread communication of written ideas. Any Process for generating creativity – human or artificial – therefore sits between two uniquely human stages: (a) a front-end of *problem definition*, and (b) a back-end of *solution validation* (Table 2.1). The focus of, and potential for, artificial creativity therefore lies between these two *human* stages. This also has a potential impact on the teaching of creativity.

2.4 *How* Are We Creative?

With the rationale for creativity established, we turn to the underlying cognitive processes. How, in other words, are creative solutions generated? Guiding this process is the fact that the novel forms may be either: (a) novel variations of what already exists, or (b) wholly new forms. The former is typically referred to as *incremental* in nature, while the latter is *radical* (Salter and Alexy 2014).

Guilford (1959) described the two essential, cognitive stages – the *processes* – of creative problem solving. These stages reflect the core issue that Guilford (1950) was seeking to redress: he argued that human intellectual ability had been characterised too narrowly as a matter of speed, accuracy and correctness, or what he described as *convergent thinking*. There was, in his view, more to human mental sovereignty than merely knowing that 2 + 2 = 4, or that the ratio of a circle's circumference to its diameter is 3.14 (pi). This is not to say that this form of cognition is trivial. Much of our lives depends on our ability to know or find, and apply, *the correct answer*. Our ability to solve problems efficiently and economically is anchored to our ability to evaluate options, to select against a list of constraints, and to calculate. All these cognitive activities are instances of convergent thinking, and in a creative,

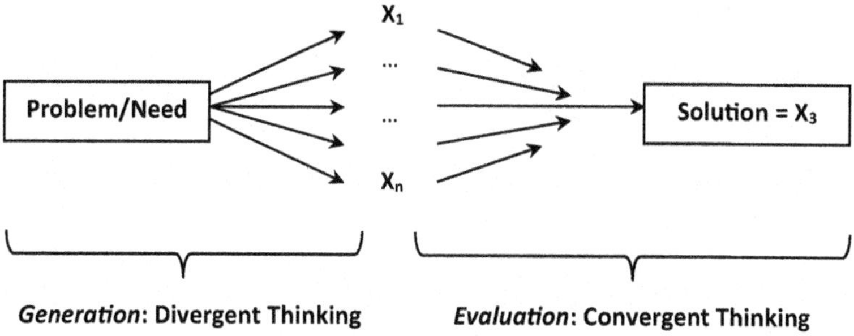

Fig. 2.2 Divergent and convergent cognition. (Based on Cropley 2015)

problem-solving process, Guilford (1959) encapsulated these under the single stage of *Idea Evaluation.*

Guilford (1950) argued that human intellectual ability was also a matter of our ability to generate alternatives, to see multiple possibilities, and to link disparate ideas. He defined *divergent thinking* – Idea Generation – as the complement to convergent thinking. In the typical course of problem solving, once a problem has been defined, a range of possible solutions must be generated through divergent thinking, before these are evaluated, and the best solution selected, using convergent thinking (Fig. 2.2).

If an artificial means of creativity is to be found – or perhaps, an artificial aid to human creativity – then it must lie in the realm of process. Can artificial systems perform idea generation, and/or idea evaluation? If so, will this obviate the need to develop these abilities in humans, fundamentally changing the relationship of creativity to education and learning?

2.5 Process & Artificial Systems

The simplest example of the core divergent process of creativity – idea generation – is *blind*, or random, generation within some *conceptual space* (Boden 2004). To illustrate this, consider the case of creativity in music. The total number of musical notes on the so-called *equal-tempered scale*, in the range of human hearing (16 Hz–16KHz), is 120. It is possible to imagine an artificial system designed to generate a novel melody (i.e., a form, or solution) based on these available notes, just as a human composer must. A sequence of four notes offers over 200 million variations (i.e., 120^4). A melody of seven notes (think of the first line of *Mary Had a Little Lamb*) is one of more than 350 trillion possibilities! Blind variation of this type is therefore capable of producing wholly new (radical) forms (melodies that have never before been heard) and might therefore be deemed *creative* (see Table 2.2).

Critical, however, is the question of effectiveness. As stated earlier, creativity is not merely the production of novel forms. As Runco and Jaeger (2012) noted,

Table 2.2 Opportunities for artificial cognition in creativity

Algorithm	Need	Potential novelty (N)	Potential effectiveness (E)	Potential creativity (N × E)	Cost of satisfying need
Blind generation	Write me a 7-note melody	Very high radical	Very low	Very low	Very high
BVSR	Write me a 7-note melody	High radical	Low	Low	Very high
Broad training	Write me a classical symphony	High incremental/low radical	Moderate	Moderate	Very high
Narrow training	Write me a symphony in the style of Mozart	Moderate incremental	High	Low	High
Plagiarism	"OK Google – Play *Eine Kleine Nachtmusik*"	None	Very high	None	Very low

"originality can be found in the word salad of a psychotic and can be produced by monkeys on word processors" (p. 92). The forms resulting from blind generation must, in the end, satisfy a need, judged by humans for their effectiveness, if they are to qualify as *creative*. In the case of music, that need might be as straightforward as eliciting a favourable reaction in the listener. Because not all combinations of notes are equally pleasing to listeners (if they were, there would be no difference between the music of Mozart, and the music of an infant banging on a toy xylophone) it is clear that this strategy for the generation of artificial musical creativity is risky, and likely to consume enormous computational resources, with no guarantee of an effective outcome. Bruner (1962a, p.6) recognised this, noting that creativity is "…not a taking of known elements and running them together by algorithm into a welter of permutations." Indeed, he went on to say that "One could design a calculator to do that, but it would be with some embarrassment, for this is stupid even for a calculator". Blind idea generation, on its own, therefore is not enough.

Even though blind generation appears to offer the possibility of high, radical novelty (i.e., wholly new melodies), the likelihood of effectiveness is very low. If creativity requires both novelty *and* effectiveness, then this method is more likely than not to produce very low creativity (see Table 2.2). In addition, the cost of attempting to satisfy the need – seeking out the *one-in-a-billion* melody that humans might deem effective – is very high.

Artificial musical creativity has, of course, progressed beyond the "stupidity" and computational inefficiency of blind generation. What is needed is an algorithm, or an *evaluative* approach, that is able to raise the likelihood of finding an *effective* melody – i.e., one that is pleasing to the listener. What is needed, therefore, is a means to raise the likelihood of successful *solution validation*. Bruner (1962a, p.7) again anticipated this, suggesting the need for "…a heuristic that guides one to fruitful combinations". Simonton (2011), echoing Campbell (1960), spoke of Blind Variation and Selective Retention (BVSR). In the simplest form, BVSR would be represented by an only slightly less stupid algorithm, designed to compute a subset of variations of notes according to some heuristic or rule, thus reducing the conceptual space. Nevertheless, this approach retains most of the same limitations of blind generation, even if it is computationally more efficient. At best, by raising the likelihood of finding an effective melody, the result of BVSR might be *low* (as opposed to *very low*) creativity.

With artificial systems designed for blind generation or BVSR *effectively* limited to low levels of creativity and high computational cost, computational creativity must turn to other means if it is to replicate, or replace, human creativity in a practical way. Can modern AI approaches such as machine learning generate novel and effective (i.e., creative) outcomes? Can machine learning replace the need for human learning of the core processes of creativity?

Unlike blind generation or BVSR of the kind described above, Artificial Intelligence (AI) and more specifically, Machine Learning (ML), offers a means to achieve a more direct pathway to Bruner's *fruitful* combinations. Take a simple case of *supervised*, classical learning.[3] An AI is *trained* on the music of Mozart. The system learns that certain combinations of notes are "correct", and is trained to generate new forms, in effect, *in the style of* Mozart. This algorithm, called *Narrow Training* here, avoids the billions of *ineffective* combinations of notes inherent in the algorithms described previously by incorporating an evaluative component. The result is more likely to be effective, precisely because it emulates music that has been defined, *a priori*, as effective. The conceptual space, in other words, is substantially reimagined. However, the stronger the training – the more the AI learns from the music of Mozart and the more it evaluates and restricts its own output – the lower the novelty. The *autonomy* (Lamb et al. 2018) of the artificial system is highly constrained. In fact, as the training becomes more focused on Mozart, what is lost first is radical novelty. The stronger the training is focused on Mozart, the more the creativity becomes incremental in nature – variations on a theme by Mozart – and if too strong, it is possible to imagine an AI unable to generate anything more than the reproduction of *Eine Kleine Nachtmusik*. The artificial system has become merely a system for plagiarising Mozart (see Table 2.2)!

Is there an approach to artificial creativity in this *idea generation/idea evaluation* paradigm that maximises the opportunity for radical novelty, while at the same time maximising the likelihood of effectiveness? The use of Machine Learning coupled with *Broad Training* has some potential. Training an AI on, for example, a set of classical composers raises the potential incremental novelty (i.e., variations on an amalgam of Mozart, Beethoven, Schubert and Haydn), and may even offer the potential for radical novelty (a new style that is a blend of Mozart, Beethoven and others). The likelihood of effectiveness is raised from that of the gamble of blind generation or BVSR, though not as high as the certainty that comes with narrow

[3]It is not our intent here to discuss, in detail, the mechanisms or methods of Machine Learning. We are drawing on well-established broad properties of different classes of Machine Learning to illustrate their potential application in artificial creativity.

training or plagiarism. The overall creativity, therefore, is potentially higher than any of the other options so far considered, though still not guaranteed. A summary of these varying possibilities is presented in Table 2.2.

There is an important caveat here. We cannot reasonably hold an artificial system to a higher standard than a human. Even human composers have bad days. This is why the *probability* or *potential* of an effective solution has to be considered, as well as the computational cost. Each approach, from blind generation to ML, can, *in theory*, meet the requirement of generating novel, and effective, outcomes (e.g., in music). Whether or not it does so remains a question for the human-centric solution validation. However, if an artificial system can only ever do as well as a human, and no better, in idea generation, and lacks autonomy, then is it a candidate to replace humans in this stage of creative problem solving? Furthermore, if an AI can only reach a modest level of creativity at a prohibitive energy cost (see Strubell et al. 2019), is the possibility of artificial creativity sustainable? Both cases suggest that the need to educate humans for creativity will remain critical, even as the impact of the Future of Work takes hold.

2.6 Case Studies in Artificial Creativity

To explore the concepts outlined, we present two case studies of artificial systems that are claimed to be, or have the apparent potential to be, creative. The first is an autonomous laboratory assistant, and the second is a story-writing AI.

2.6.1 The Catalysing Computer

In mid-2020, researchers at the University of Liverpool in the UK (Burger et al. 2020) announced the creation of an intelligent, mobile, robot scientist, capable of undertaking chemical experiments independently of humans.[4] The robot in question works in a normal laboratory, using typical *human* instruments. The Liverpool robot is able to work for up to 21.6 hours each day, stopping only to recharge its batteries. In Burger et al's study, the robot performed 688 experiments in 8 days, and found new photocatalyst mixtures. The researchers estimated that the same experiments would have taken a human several months to perform. The robot scientist used a *batched Bayesian search algorithm* (see, for example, Shahriari et al. 2015) to explore a 10-dimensional search space, with more than 98 million points. This technique, broadly speaking, is a form of supervised machine learning. As such, the University of Liverpool's robot scientist is undoubtably an exemplar of a robot taking over the work of a human. However, is that work routine and algorithmic –

[4]See: https://www.eurekalert.org/pub_releases/2020-07/uol-lrb070620.php

with the robot simply doing it faster and more accurately than a human – or is the work creative?

Using the criteria listed in Table 2.2, it seems clear that this is an example of *narrow training*. The system was programmed, very explicitly, to identify photocatalyst mixtures with particular properties. Indeed, the robot scientist did uncover photocatalyst mixtures "that were six times more active than initial formulations" (Burger et al. 2020, p. 237) indicating not only a clear capacity for some degree of incremental novelty, but for *effective* incremental novelty. However, it was trained, so to speak, for this one very specific purpose. In the same way that we would not expect the music AI trained on Mozart to suddenly compose a piece in the style of Stravinsky, the robot scientist could not autonomously decide to pursue different experiments. It could only ever (without human intervention, in the form of reprogramming) search for photocatalysts. Indeed, unless the laboratory was restocked by a compliant human, the robot scientist could not find anything at all. The robot scientist lacks autonomy, with the authors noting (p. 241) that "this autonomous system does not at present generate and test scientific hypotheses by itself".

Summarising the abilities of the robot scientist against the criteria in Table 2.2 suggests that it possesses a limited capacity to *support* incremental creativity. It cannot autonomously decide what need is to be satisfied, and it cannot decide if its output actually satisfies that need. Those decisions – *problem definition* and *solution validation* – remain a matter for the humans who program and control the robot scientist. It can, however, explore a large, specific conceptual space much more quickly than a human scientist, evaluating candidate solutions, and presenting those to its human colleagues in a form of highly efficient *technology push* (Martin 1994; Cropley and Cropley 2015). In this sense, perhaps artificial intelligence should be thought of as a support in the creative process even, as one team member noted, "free [ing] up time for the human researchers to think creatively".[4] Critically, the need to educate humans in all stages of the creative problem-solving process remains unchanged. Not only to understand how to direct and make use of the abilities of the robot scientist, but to bridge the gap between supporting incremental creativity, and achieving radical creativity.

2.6.2 The Android Author

In 2016, the Kimagure Artificial Intelligence Writer Project, based in Japan, produced a story,[5] which was submitted as an entry for the Nikkei Hoshi Shinichi

---

[5] https://www.smithsonianmag.com/smart-news/ai-written-novella-almost-won-literary-prize-180958577/

Literary Award.[6] The story – *The Day a Computer Wrote a Novel* – was hailed for the fact that it made it through the first of four rounds of evaluation by the competition's (human) judges. It was eliminated in the second round.

The process that delivered this apparently creative output was, however, not as straightforward as an AI *writing a novel.* A story was first written by the team responsible for the project. This was then *structuralized*: it was broken into its components parts (words). In another example of classical, supervised learning, the researcher team then fed these data into their AI, along with a set of instructions/rules for how to reassemble the words into a coherent story. Perhaps not surprisingly, the result was something resembling a story.

Assuming that the algorithm was some form of machine learning, and not merely pseudo-random recombination of the original words, then this example is at best an instance of *narrow training*, possibly bordering on a form of automated self-plagiarism. It is unclear if the original story served as an exemplar of a good story from which the AI could learn, or merely a library of possible words, comprising a finite conceptual space. Regardless, we can apply the criteria of Table 2.2 to assess the potential creativity of this story-writing AI.

With a single exemplar used to train this AI, and a highly constraining set of rules for reassembling the story, the potential novelty of this example remains frustratingly incremental. Unlike the robot scientist, the incremental novelty here is not simply an output that is "two times faster" or "three times more efficient", but incremental in the sense of "what happens if the main character finds a new object in chapter 2, instead of losing something they had?" The novelty is reflected in improvements to, or changes in, a baseline output.

The story-writing AI did produce an output representing at least some degree of effectiveness. Unlike the robot scientist, for which effectiveness is more explicitly definable, the output of the story-writing AI is judged with a greater degree of subjectivity. However, this does not mean that effectiveness is undefinable for an "artistic" output – only that it may be subject to greater variability. However, as the Consensual Assessment Technique (see Amabile 1982, 1996) has shown, it is perfectly possible for knowledgeable experts to reach agreement, even for outputs of an artistic nature. Therefore, while you may like *The Day a Computer Wrote a Novel*, the expert judges, in the end, did not.

Consequently, with only moderate incremental novelty, and even with high effectiveness, the story-writing AI is, like the robot scientist, limited to *supporting incremental creativity*. The need, in this case defined not only by the original (human-authored) story, but also by the rules for writing the AI-authored story, remains firmly the job of humans. Solution validation, similarly, remains a task for humans. Moreover, like the case of the robot scientist, the need to educate humans in all facets of creativity and creative problem solving has not been removed.

[6]The Nikkei Hoshi Shinichi Literary Award is open to both human and non-human (i.e., AI) authors.

2.7 AI and Humans in Partnership

The majority of work conducted in organizations is done in teams: especially creative teams (Reiter-Palmon et al. 2012). In addition, to the subject of how AI can be creative, it is paramount to consider the partnership of AI and human integration in solving creative problems. Creative teams typically benefit from varied perspectives, multiple domains of expertise, and increased knowledge. Integrating AI into this formula may serve to free-up some cognitive stress placed on humans (i.e., acquiring resources in terms of information). Although AI lacks conceptual thinking, developing strategies that facilitate AI and humans working together, boasting the benefits of AI (i.e., rapid information acquisition) may allow for *more* creative efforts by addressing areas that typical human-centered teams struggle with. In simple terms, AI can serve as a method to implement checks and balances to circumvent weak spots within teams.

Despite the attraction of integrating AI into teams, there are, of course, potential drawbacks. High-performing teams balance on a high level of psychological safety and trust (Braun et al. 2013; Lee et al. 2010; Langfred 2004). Thus, one should consider how incorporating AI into teams impacts the composition and configurations of the teams. Although AI can aid creative teams by having the ability to collect, store, and process information at high speeds (Elkins and Derrick 2013), these benefits are fruitless if that same information is not shared due to a lack of trust in using AI. Notably, humans have an initial distrust from their beginning interaction with AI (Sheridan and Hennessy 1984). That said, more interaction with AI systems and their popularity in mainstream press appearances may increase the confidence a human places on the benefits and likelihood of using the AI system (Ashton 1990). In evidence of this, Elkins and Derrick (2013) demonstrated that the trust within an interaction between humans and AI was temporally variant and could be explained by a linear change in time.

Taken together, the avenue on integrating a partnership of AI and Humans may be better positioned from the focus of *Humans* and AI. Specifically, attention to how AI will influence teamwork and strategies that facilitate this experience will be vital for organizational performance.

2.8 Conclusions

The question *can robots be creative* grows in significance the deeper we move into the era of Industry 4.0. If only humans can generate novel and effective ideas, then the Future of Work holds no fears for us, provided we educate ourselves to be effective creative problem solvers. If, on the other hand, robots displace humans even in this key ability, then we face significant challenges. Indeed, we face the prospect of being confined to the jobs that *even the robots refuse to do*[7]!

In this chapter we have hypothesised that with regard to the defining activity of creativity – the ability to generate novel and effective outputs – artificial systems are limited to, at best, moderate levels of incremental creativity. In other words, artificial systems have the potential to generate new, and effective, variations of existing ideas, solutions, systems, and artefacts. Furthermore, even with improvements and changes to the technology of AI, this capacity is likely *not* to transition, eventually, into an autonomous ability for radical creativity, but simply into higher levels of incremental creativity, at a lower cost. While computing power will increase, and algorithms will continue to improve, the limiting factor on artificial creativity (aside from possible data and energy constraints) is not *how* (the process): it is the rationale (why). No matter how good AI technology becomes, the reason why we are creative – problem definition and solution validation – remains the job of humans. Furthermore, the importance of educating humans for creativity becomes stronger as the jobs of the future coalesce around non-routine, non-algorithmic, non-automatable jobs and tasks.

Nevertheless, the debate will no doubt continue. It is driven, however, not by questions of technology, but by deeply ingrained myths and misconceptions about creativity that continue to dog this field of research (e.g. Patston et al. 2018). The discipline of psychology has been consistent in its definition of creativity – always anchored to the production of novel and effective outcomes – however, the computational field has exhibited a much greater flexibility. Newell et al. (1962) were closely aligned with what psychology now refers to as the *standard definition* (Runco and Jaeger 2012), however, as Wiggins (2006, p. 210) demonstrates, computational definitions also frequently choose to focus on "…behaviour which would be deemed creative if exhibited by humans." Embedded in the Wiggins definition is an important, yet false, premise. The human behaviour referred to seems to be the *act of creating* – in other words, the assumption is that "humans bring things into being [they create], and computers can be programmed to bring things into being [to create], therefore, computers are *creative*." This ignores the standard definition of creativity, as it ignores the rationale for creativity.

Taking this into account, we are left with one final consideration. What value is derived from the ability that artificial systems do possess to *support* the core processes of creativity: idea generation and idea evaluation? As the cases of the *Catalyzing Computer* and the *Android Author* illustrate, artificial systems first and foremost dramatically speed up tasks that are necessary, but not sufficient elements in the process of creative problem solving. They relieve humans of slow, repetitive, algorithmic tasks – whether trying out hundreds of possible chemical formulae or exploring the potential of different story lines – creating more time and more

[7] This could be either in the sense that robots, literally, pick and choose what they do, or figuratively, in the sense that the owners of expensive, intelligent robots would not waste this resource on menial, dangerous or dirty tasks that risk damaging their robot.

opportunity for humans to identify unresolved needs and match these to genuinely new and effective solutions.

References

Amabile, T.M. (1982). Social psychology of creativity: A consensual assessment technique. *Journal of Personality and Social Psychology* 43, 997–1013.
Amabile, T.M. (1996). *Creativity in context.* Boulder, CO: Westview Press.
Ashton, R.H. (1990). Pressure and performance in accounting decision settings: Paradoxical effects of incentives, feedback, and justification. *Journal of Accounting Research* 28, 148–180.
Atlantic, T. (2018). Available: https://www.theatlantic.com/technology/archive/2018/10/automated-on-the-road/571345/ [Accessed].
Barron, F.X. (1969). *Creative person and creative process.* New York, NY: Holt, Rinehart & Winston.
BBC (2017). Available: https://www.bbc.co.uk/news/technology-40042581 [Accessed February 23, 2021].
Boden, M.A. (2004). *The creative mind: Myths and mechanisms.* Hove, UK: Psychology Press.
Braun, S., Peus, C., Weisweiler, S., and Frey, D. (2013). Transformational leadership, job satisfaction, and team performance: A multilevel mediation model of trust. *The Leadership Quarterly* 24, 270–283.
Bruner, J.S. (1962a). "The Conditions of Creativity," in *Contemporary Approaches to Cognition,* eds. H. Gruber, G. Terrell & M. Wertheimer. (New York, NY: Atherton Press), 1–30.
Bruner, J.S. (1962b). The New Educational Technology. *The American Behavioral Scientist* 6, 5–7.
Burger, B., Maffettone, P.M., Gusev, V.V., Aitchison, C.M., Bai, Y., Wang, X., Li, X., Alston, B. M., Li, B., and Clowes, R. (2020). A mobile robotic chemist. *Nature* 583, 237–241.
Campbell, D.T. (1960). Blind variation and selective retentions in creative thought as in other knowledge processes. *Psychological Review* 67, 380–400.
Cropley, D. H. (2015). *Creativity in engineering: Novel solutions to complex problems.* San Diego: Academic.
Cropley, D. H. (2019). *Homo Problematis Solvendis - Problem-solving Man: A History of Human Creativity.* Singapore: Springer Nature.
Cropley, D. H. (2020). *Femina Problematis Solvendis - Problem Solving Woman: A History of the Creativity of Women.* Singapore: Springer Nature.
Cropley, D.H., and Cropley, A.J. (2015). *The psychology of innovation in organizations.* New York, NY: Cambridge University Press.
Elkins, A.C., and Derrick, D.C. (2013). The sound of trust: Voice as a measurement of trust during interactions with embodied conversational agents. *Group Decision and Negotiation* 22, 897–913.
Guckelsberger, C., Salge, C., and Colton, S. (2017). "Addressing the "why?" in computational creativity: A non-anthropocentric, minimal model of intentional creative agency", in: *8th International Conference on Computational Creativity.* (Atlanta, GA: Association for Computational Creativity).
Guilford, J. P. (1950). Creativity. *American Psychologist, 5*, 444–454.
Guilford, J. P. (1959). Traits of creativity. In H. H. Anderson (Ed.), *Creativity and its cultivation* (pp. 142–161). New York, NY: Harper.
Lamb, C., Brown, D. G., & Clarke, C. L. (2018). Evaluating computational creativity: An interdisciplinary tutorial. *ACM Computing Surveys (CSUR), 51*(2), 1–34.
Langfred, C.W. (2004). Too much of a good thing? Negative effects of high trust and individual autonomy in self-managing teams. *Academy of Management Journal* 47, 385–399.
Lee, P., Gillespie, N., Mann, L., and Wearing, A. (2010). Leadership and trust: Their effect on knowledge sharing and team performance. Management Learning. *41*, 473–491.

Martin, M.J.C. (1994). *Managing Innovation and Entrepreneurship in Technology-based Firms.* New York, NY: John Wiley & Sons.
Newell, A., Shaw, J.C., and Simon, H.A. (1962). "The processes of creative thinking," in *Contemporary Approaches to Creative Thinking,* eds. H. E. Gruber, G. Terrell & M. Wertheimer. (New York, NY: Atherton Press), 63–119.
OECD. (2017). *Future of work and skills.* Retrieved from: https://www.oecd.org/els/emp/wcms_556984.pdf
Patston, T., J., Kaufman, J.C., Cropley, A.J., and Marrone, R.L. (2021). What Is Creativity in Education? A Qualitative Study of International Curricula. *Journal of Advanced Academics.*
Patston, T.P., Cropley, D.H., Marrone, R.L., and Kaufman, J.C. (2018). Teacher Implicit Beliefs of Creativity: Is there an Arts Bias? *Teaching and Teacher Education* 75, 366–374.
Reiter-Palmon, R., Wigert, B., and De Vreede, T. (2012). "Team creativity and innovation: The effect of group composition, social processes, and cognition," in *Handbook of organizational creativity.* Elsevier), 295–326.
Rhodes, M. (1961). An analysis of creativity. *The Phi Delta Kappan, 42*(7), 305–310.
Runco, M.A., and Jaeger, G.J. (2012). The Standard Definition of Creativity. *Creativity Research Journal* 24, 92–96.
Salter, A., and Alexy, O. (2014). "The Nature of Innovation," in *The Oxford Handbook of Innovation Management,* eds. M. Dodgson, D.M. Gann & N. Phillips. (Oxford, UK: Oxford University Press), 26–49.
Shahriari, B., Swersky, K., Wang, Z., Adams, R.P., and De Freitas, N. (2015). Taking the human out of the loop: A review of Bayesian optimization. *Proceedings of the IEEE* 104, 148–175.
Sheridan, T.B., and Hennessy, R.T. (1984). "Research and modeling of supervisory control behavior. Report of a workshop." (Washington DC).
Simonton, D.K. (2011). Creativity and discovery as blind variation and selective retention: Multiple-variant definition and blind-sighted integration. *Psychology of Aesthetics, Creativity, and the Arts* 5, 222–228.
Strubell, E., Ganesh, A., and Mccallum, A. (2019). "Energy and policy considerations for deep learning in NLP", in: *arXiv preprint arXiv:1906.02243.*)
WEF (2016). "The future of jobs: Employment, skills and workforce strategy for the fourth industrial revolution", in: *Global Challenge Insight Report.* World Economic Forum).
WEF (2020). "The Future of Jobs Report 2020").
Wiggins, G.A. (2006). Searching for computational creativity. *New Generation Computing* 24, 209–222.

Dr. David H. Cropley is an international author, researcher and consultant in creativity and innovation, specialising in helping people and organisations to unlock their idea generation and problem-solving capabilities. David's expertise is based on a deep understanding of the psychology of creativity and innovation, coupled with real-world experience in technology and engineering. David has worked with organisations in North America, Europe, Asia and Australia, guiding them through the complexities of value creation in an ever-changing and challenging world. He is currently the Professor of Engineering Innovation at the University of South Australia in Adelaide.

Dr. Kelsey E. Medeiros is an Assistant Professor of Management at the University of Nebraska at Omaha. Her research focuses on "troublemaking," including creativity and innovation, (un)ethical behavior, sensemaking, women in leadership, and sexual harassment.

Dr. Adam Damadzic is an Industrial and Organizational Psychology (I/O) graduate of the University of Nebraska at Omaha. He completed his masters in I/O Psychology at the University of Texas at Arlington. His research interests include creativity & innovation, leadership, and gender inequality in organizations. Additionally, Adam is a Senior I/O Psychologist for T-Mobile. Adam has published and has applied experiences in the areas of individual creativity, teamwork, and leadership.

Chapter 3
Conceiving Creativity and Learning in a World of Artificial Intelligence: A Thinking Model

Edwin Creely

Abstract In this chapter I engage with the evolving issue of how artificial intelligence (AI) relates to human creativity and the present and future possibilities of how AI might change the ways creativity is understood and applied. The chapter begins with the survey of the territory, historically and conceptually, to establish the ground for the discussion, with focus on understandings of creativity, consciousness, learning, and intelligence. The problem of defining these terms in the light of the development of AI is discussed. Then I propose a thinking model for the intersections between creativity and AI containing four distinct spaces of intersection built on the notions of 'narrow' and 'wide' as pertaining to both creativity and AI. The purpose of the thinking model is not to offer any definitive answers but to suggest areas for exploration and to foster conversations in this emerging area of research and practice. Practical examples and implications that emerge from my exploration of the model are offered, with particular focus on examples related to education.

Keywords Creativity · Artificial intelligence (AI) · Consciousness · Intelligence · Learning · Thinking model

3.1 Introduction

Artificial Intelligence (AI) is becoming a buzz word in many contexts around the world and is being seen as a strategic component of so-called futures (Dator 2020; Köstler and Ossewaarde 2021). In education and in schools, which exist in an increasingly digitised world, AI is looming on the horizon in its potential impact on teaching and learning (Bozkurt et al. 2021). The pervasiveness of AI globally across all sectors and industries is now a given when considering strategic planning and policy implementation in education. At the same time, its potential uses in education are still somewhat elusive or inconclusive, and many questions, both

E. Creely (✉)
Faculty of Education, Monash University, Melbourne, Australia
e-mail: edwin.creely@monash.edu

D. Henriksen, P. Mishra (eds.), *Creative Provocations: Speculations on the Future of Creativity, Technology & Learning*, Creativity Theory and Action in Education 7,
https://doi.org/10.1007/978-3-031-14549-0_3

practical, ethical, and theoretical, remain. AI is also being viewed as a mechanism of facilitating creativity and enhancing creative processes and products (de Vasconcellos et al. 2021). But how exactly does AI work with creativity? This is an emerging question.

This chapter explores the notion of AI (in its narrow and general usages), and what it might provide in terms of promoting creativity in educational settings with technologies. The chapter critically investigates the ways that AI might support the creativity of both teachers and students, and the possibilities that emerge in AI-human interactions in education. The chapter is grounded in the theoretical perspective of post-humanism and presents provocations around what the human-machine interactions might mean for a re-imagining of creativity in education. The disposition of creativity remains an area of dispute in scholarship, and its links to AI even more so. The meaning of this term is explored in detail later in this chapter, but for now creativity is taken to mean novel processes and products that emerge from and are embodied in the work of individuals, groups, and systems (Csikszentmihalyi 1997, 2014; Creely et al. 2021).

To investigate the AI-creativity relationship, one only beginning to be teased out, with no apparent definitiveness, I decided to offer a *thinking model* to work through some of the potential interconnections and possibilities. There are many thinking models that have been developed as part of adopting a 'flowing' thinking process (Kalyta and Klymenyuk 2018). One of the most popular is that by Tim Hurson (2007) in which a strategic thinking process is unfurled aimed at a specific problem, which leads to a solution. Since the potential problems with AI and creativity are only just emerging and there are no ready solutions to be had, my thinking model is oriented to possibilities, conjectures, and provocations—about initiating some thinking and even sparking disagreement in an open-ended way. I am mapping, typologically, some of the territory, but it may well resist being revealed.

3.2 The Possibilities in AI

From its beginnings as a research project at Dartmouth College in the 1950s to recent nuanced applications, such as big data analytics, speech and language recognition (particularly in Siri, Cortana and Google Translate etc.), facial recognition, forecasting, online advertising, moderation of social media, robotics, augmented and virtual reality (AR, VR), driverless transport, navigation systems, solving complex problems, and surveillance (and the list goes on), AI has been enmeshed with the growth in and capabilities of computer technologies (Boden 2018; Mitchell 2019; Verma et al. 2021; Yi and Loia 2019). The development of data centres in the last decade (Google and Apple being most notable) has fostered a distributed approach to computing and the handling of data that enables the narrow application of AI to specific problems and specialized applications, as well as for everyday consumer use. Put simply, starting from its tentative beginnings at the outset of the so-called computer revolution in the 1950s, AI is emerging rapidly and converging with all areas of human life and endeavour, including in creative output (Lee and Trimi 2021).

With the increased processing power of computers, and specialised processing within CPUs that include neural engines, comes (visionary) projections about where AI might go, especially in moving beyond narrow conceptions of AI based on algorithms and machine learning to ones involving general and super intelligence that are human-like, brain-like (with complex neural networks). The promises for AI in the computation power of quantum computing appears to be opening new horizons about what might be possible. These projections are conceived in the popular imagination often in futuristic terms with applications in robotics, autonomous machines that are capable of unsupervised deep learning and the prospect of new non-biological lifeforms that may supersede human intelligence (Bunz and Braghieri 2021).

There is even the fanciful goal of the emergence of an entity such a Data from Star Trek Next Generation, a non-carbon life form capable of self-awareness, consciousness, spirituality, and the ability to learn independently of explicit human training, and to be creative in the sense of independently forming new and unexpected insights and outcomes (Andrews 2015). It is easy to dismiss this imagined goal and the notion of 'the singularity is near' as espoused in 2005 by Raymond Kurzweil of Google (Kurzweil 2005). However, it remains an objective for some AI researchers committed to research on complex neural networks, who envision a posthuman evolutionary future that they are convinced will inevitably come (Brockman 2020; Naudé 2021).

So, there are two distinct possibilities that might be imagined for AI now and into the future, especially regarding the intersections of technologies and machines with humans and their creativity. The first is for an AI that *non-obtrusively occupies the support spaces* around human lives in online applications, information and navigation systems, social media, online shopping and selling, voice applications, face recognition, communication systems and the storage, retrieval, and analysis of data. It is now also pivotal to the work of global corporate entities, especially in data analytics and market predictions. This non-intrusive use of AI has become more sophisticated and is positioned with virtually every use humans now make of technologies. Such an AI sits beside and mediates the work, leisure, and communication experiences of billions of humans on this planet and is likely to become even more pervasive than it is now (Yolgörmez 2021). But it is essentially narrow in its scope, relying on machine learning, sophisticated algorithms, and distributed computing networks. One example for this use of deep machine learning is the concept of evolutionary algorithms that mimic the process of biological natural selection to solve hard problems through a series of mutations and crossovers into new forms, with the best form the one that survives and solves the problem (Soni 2018).

In terms of creative processes and outcomes, this possibility for AI is about a *supplementary role* in human creativity and learning or an augmentation of what humans currently do. Digital AI tools, including ones in VR and AR, provide scope to be creative in new ways, built on the processing power to increase the pace of creative production as well as an immersive space for working with productive failure and iterative processes in creative practices (Mikalef and Gupta 2021). This may have an impact in teaching and learning environments and instructional delivery, not just in the arts but across discipline areas in schools and higher education institutions (Popenici and Kerr 2017).

The second possibility is an emerging AI that is *disruptive* to the ways that humans conceive of their place in knowledge making and creative output. This disruptive possibility for AI sits squarely in the realm of deep learning and research on sophisticated neural networks that embody technologies that go beyond narrow uses of algorithms and supervised deep learning systems. In this possibility, that is only just beginning to emerge, complex super AI systems that match (and may theoretically go beyond) human intelligence create knowledge and manage decision-making alongside and even outside human control (Peeters et al. 2021).

In the context of this super AI and the disruption of the centrality of human futures and technological advancements, creativity might move past an anthropomorphic perspective to a posthuman positionality in which human intelligence and AI sit inexorably together (Anderson and Rainie 2018). Discovery, innovation, and development of ideas become more than a human enterprise as non-human super AI emerges with creative agency. Human learning, in this scenario, is in a nexus with non-human and material agencies, where there is the possibility of learning that subverts the usual role of teacher or instructor, for instance.

3.3 AI and Creativity: Contested Territory

AI (including machine learning and deep learning categories) and its expansion over the past 70 years, with significant acceleration of development in the last 10 years, is not without its controversies and remains a contested territory of technological innovation and ethical consideration. In unfolding my thinking model, these areas of dispute and contestation are important to explicate. In terms of AI and creativity, this contestation is evident in three areas.

Intelligence The first is about the use of the term 'intelligence' as an operative word to frame the territory of innovation and research to do with AI. This term has no fully agreed definition and there are diverse ways of conceiving intelligence (Hawkins and Blakeslee 2004). Some understandings of intelligence are about discrete cognitive processes and behaviours, while others account for a diverse range of human capacities, include specialised and disciplinary specific intelligences (Schlinger 2003; Staats and Burns 1981). In broad terms intelligence could be characterised as a set of capacities to effectively engage in the world, adapt to new circumstances and solve theoretical and practical problems. Sternberg's theory of intelligence includes three discrete types: analytical, practical, and creative (Sternberg et al. 2008). This theory conceives of intelligence as modes of operation of humans interacting with the world through their subjectivities. Thus, it is not unreasonable to suggest that intelligence is more than cognition and intellectual capabilities and involves diverse ways of embodying in the world, and this holistic sense of intelligence is observable in a range of modalities, schema, and subjectivities (Morasso and Mohan 2021). However, this holistic notion of intelligence is debatable, especially about whether conceiving intelligence as a set of operational modes and schemas of engaging with the world is consistent with psychological research and brain studies (Sternberg 2020).

Regarding AI, intelligence and creativity, there appears to be an underlying assumption about what intelligence is, which usually centres on 'mental' processing that is quite apart from the body and embeddedness in the world. For narrow conceptualisations of AI, this view of intelligence is about the completion of discrete tasks, for example, speech recognition and synthesis, and more-or-less emulates delimited human cognitive functionality. The replication of human intelligence at a general AI level becomes more problematic since it about virtual mimicry of complex human brain capacities, including thinking, intricate problem-solving, responding, awareness, decision-making, predicting and the like. The contestation in terms of creativity is about not only the extent to which AI can replicate or mimic complex human cognitive functionalities, including divergence, but also about the disposition of what it means to be creatively intelligent as an autonomous being (Varghese 2021). Because there are disputed understandings of what it means to be creative, the notion of developing creatively intelligent AI, alongside human creative functionality, is possibly elusive (Rust and Huang 2021).

Consciousness Contestation is also apparent when considering the parallel notion of consciousness. What consciousness is and what creates it has been disputed in a range of discipline areas, including brain science, even before the emergence of AI research, and debates about consciousness go back to the classical philosophers (Sattin et al. 2021). Whilst acknowledging that biological creatures have it, and arguably humans have it more with their greater brain capacity, how to understand and define it remains tricky, as does its connections with other characteristics related to the human mind, the body and intelligence (Shear 1997). It might be viewed as the *ground* for or the underlying state of being alive (beingness) in the world upon which perception, awareness, feeling states, embodying, and thinking, as well as memories, histories, communications, and adaptabilities are built—a sort of null underlying state of being biologically switched on. It is a state experienced and evident in the presence of a person that is alive. Having written all this, I am also aware that this notion of a null state or 'aliveness' to describe consciousness is rather vague and lacks specificity (indeed I wish I could do better). Then again, describing consciousness as brain function has the potential to be reductionist regarding what is a complicated construct at the heart of human experience. All this, of course, remains an open question.

The goal of general AI, and even more so for strong or super AI, is movement towards an independent or unsupervised non-biological intelligence that also encompasses consciousness—a non-biological state of aliveness and presence or complex (multi) neural-like functionality. Having such a goal is admirable, but the practicalities and specificities of achieving this goal appear insurmountable in the circumstance where consciousness is not really understood and there are no fully agreed definitions or understandings about what it is (Li et al. 2021). This goal may also involve complex ethical issues about the experiences (even the potential suffering) of such proto-sentient, non-biological beings (Bell 2018; Klein 2021).

In consideration of creativity this inexorable state of being alive (of consciousness) seems pivotal to the designation of what is creative and what is not merely a category in language. A key question emerges in this debate about the link between human consciousness and creativity: Is creativity an especially human quality and linked to language, and only realised in consciousness of it? If one takes this anthropomorphic perspective then a process and product cannot be deemed creative unless it is primarily driven by human cognitive, embodied, and affective processes, and then affirmed as creative in the language constructs that mediate all our understandings of participating in the world and with each other.

Junctures with Human Learning A third area of contestation is about the junctures of AI with human learning and within the circumstances in which learners are creative (Horvatić and Lipic 2021). The generic concept of AI has captured the public's imagination as this yet-to-come wave from the future that is as much a threat as it is a blessing. There is contestation around the public discourse about AI and its place in learning and education and what educational futures might look like. AI is being touted by some as core to the future of education.

If learning is about the development of both content knowledge and skills formation in the matrix of a changing digital world that requires adaptability, then AI is likely to have, and is already having, a profound impact on teaching and learning within educational settings (Navidi and Landry 2021). Many questions emerge from consideration of impact. Not the least of these questions is about the locus of control of learning in an AI mediated set of possible futures. What will be the role of the teacher in such futures, and how will this dynamically play out in *how* students learn, or even *where* they learn? It is not my wish to answer this complex question other than identifying the space of contestation and tensions, which also includes ethical considerations regarding the global issue of the control that tech companies might have over AI applications for learning. Perhaps, emerging out of the thinking model explicated below, some of these possible areas for contestation and tension will be suggested as the spaces for thinking are opened.

There is also the question, central to this chapter, about the junctures of AI with learning and the potential for creativity of students. AI within digital and online spaces provides the possibility for the design and construction of 'safe spaces' where students can take risks and be exploratory in their learning, and not averse to failure. As discussed above, AI takes a supplementary role in sponsoring creative learning and opening possibilities and avenues in which students can be creative. But there is also the emerging likelihood of disruption to what is normally considered human processes and outcomes that are creative. As AI becomes more human-like, the locus of control within student creativity may well shift as the machine itself becomes creative.

3.4 Adopting a Posthuman Perspective

Which brings me to the question of the apparent anthropocentric bias of considerations about both learning and creativity. Posthuman and new materialist theories of knowing point to the inevitable place of the material and the non-human in considerations of futures and the disposition of learning and creativity (Mellamphy 2020). These theories give greater credence to the agency of non-human entities and to the shaping-effect of such agency on human learning and productivity (Höppner 2021). Such theories also suggest the possibility of the autonomous emergence of AI as capable of novelty and complex creative outputs. If this approach to epistemology is taken seriously, then AI and human creativity might be seen in innovative juxtaposition—with AI challenging the boundaries of what is taken to be creative and who is deemed to be creative.

In this sense, AI and the emergence of the autonomous creative machine might be viewed as another momentous step in human evolution and the intersection of biological and non-biological awareness and consciousness. As such, AI becomes an evolutionary invention to shift the human species to another evolutionary epoch. According to Max Tegmark (2017), this is Life 3.0, where non-biological intelligent life has the capacity to think, create, be conscious and fully autonomous, living in equivalent status with intelligent biological life, and perhaps even determining the direction of future evolution.

Of course, this idea of the fully creative AI machine is purely speculative at this stage. But it does evoke the question about the significant intersectionality of the material (including AI) with human creativity and learning and problematizes the notion of the human being positioned above the natural and material world. It also challenges the belief about the solitary creative individual human who exists as a thinking/creating creature quite apart from the world and the materiality of the world. Emphatically, human creativity is contingent upon the affordances and limitations of what the material provides, and further, this materiality has a profoundly agential effect on both the boundaries in which creativity can emerge and the content of human learning and creative output. Both meaning and content in creativity are found in the nexus between human agency and creativity and the agency of the material itself.

3.5 Intersections Between Creativity and AI

The previous discussion in this chapter points to the complicatedness and murky definitional territory of AI, intelligence, consciousness and how this plays out with creativity and learning. I also engaged with ideas from posthuman scholarship to provoke conjecture about how AI and creativity might be reimagined in terms of non-human entities and in processes in which human agency might not be primary—the taking of non-human agencies seriously in considering creativity and learning.

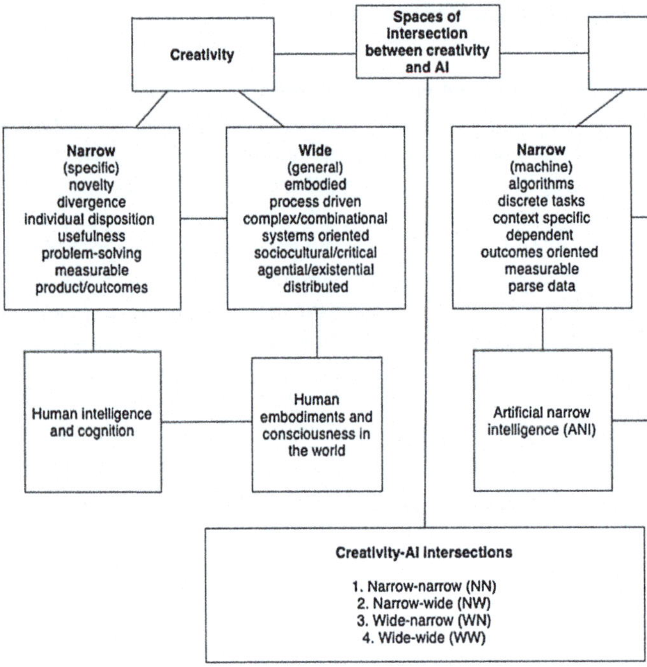

Fig. 3.1 A thinking model of intersections between AI and creativity

In this section I offer a model (Fig. 3.1) for reflecting on the nature of creativity in nexus with AI and point to quite specific ways in which creativity and AI might interact in learning settings. This model might be viewed as a thinking model for unfolding possibilities and futures—and instigating argument and provocations. As such, it should not be seen as conclusive but exploratory in examining intersections that already exist and ones that are yet to come.

The goal of this thinking model is to bring attention to the spaces of potential intersection between AI and creativity (Sanders and Wood 2021). This might lead to ways of conceiving how creativity and AI might be positioned together for creative learning or perhaps how they work in synchrony in educational contexts. There is also the possibility of emergent difficulties and issues within each of these spaces that should be accounted for.

One foundational idea for the formation of the model is to conceive both creativity and AI as essentially narrow and wide in scope for both research and practice. In the diagram, the line joining the narrow and wide set of characteristics for creativity and AI is meant to indicate that there is a continuum and so these ideas are not positioned in a binary construction. The narrow and wide (or general) understandings of AI have been well documented in academic and popular literature and the distinction between the two is now quite understood, though the rapid developments in the field may shift these lines of understanding.

This is less the case in terms of creativity. In this model, I understand creativity in its narrow designation as having a focus especially on individual cognition, divergence, and with implied measurable outcomes (Moruzzi 2021); and wide in which creativity is seen through the lens of complexity and systems-based understandings, along with the notion that creativity is socially and culturally contextualized, embodied in contexts, collaborative and distributed, and involves consciousness (Bown 2012; Clarke and Doffman 2017; Csikszentmihalyi 2014). For the wide view of creativity there is also an emphasis on agency and the existential foundations of human creative initiatives in concert with the organisation and the group. Creativity in its wide sense is also more than 'thinking outside the box', for it includes self-expression, critical engagement, and novel ways of embodying and doing.

However, both the narrow and the wide view of creativity should be seen as inextricably connected and 'narrow' and 'wide' become arbitrary points of emphasis in research and practice rather than rigid points of differentiation.

Given this mechanism for understanding both AI and creativity, I postulate a set of four intersectional spaces that represent combinatorial possibilities between creativity and AI as both narrow and wide. They are, of course, not the only combinatorial possibilities. For each of these combinations I also offer two examples of specific applications, with awareness that there are, of course, many more. These examples function as a thinking catalyst to conceive of further applications and possibilities.

(i) **Narrow (creativity)/Narrow (AI) (NN)**

In this intersectional space, creativity is a highly delineated and individual characteristic associated with cognition and seen in measurable outcomes, divergent thinking, and problem-solving capabilities. AI can both mimic and extend human narrow-creativity in highly focused problems and specialized applications, in a limited range of functionality. In this space, algorithms can be complementary to human creativity with a restricted task orientation or specificity of application. Two examples might include:

- *Data mining* where the interconnections within and between data sets are made to solve problems, in intersection with human problem-solving activities and novel solutions. In educational contexts this could be part of informed decisions about trends that require an innovative response to manage change.
- *Language processing* to support translation, transcription, spelling and syntax correction, and language learning in diverse language environments such as those often seen in classrooms where applications support the uses of language by learners and to provide a resource for their creative language-rich endeavours (Taulli 2021).

(ii) **Narrow (creativity)-Wide (AI) (NW)**

In this intersectional space, creativity is conceived as having a narrow, conceptual focus and oriented to specific problems and finding novel solutions in circumscribed settings or narrow parameters. AI, in this space, has a more expansive role—more human-like—in bringing high levels of prediction and the ability to adapt within complex structures, control systems and multifactorial problems. In this the AI offers virtual human reasoning and understandings that account for the intricacy of circumstances.

Two examples in this space that have application to education might be:

- *Sophisticated 'deep' natural speech processing* and interactions between machine and human that account for the nuances of culture, accent, regional variations, and context. This process is about information gathering, multimodal communication and complex control systems in which the AI can act independently and provide decision-making input. The beginnings of this application are unfolding for the public in voice assisted communication systems such as Alexa, Siri and Cortana that are increasingly employing AI for its predictive potential in communications.
- *The construction of deeply immersive 3D communication and learning environments* that have a natural sense of a physical environment, with rendered objects and realistic bots that represent humans interacting virtually. Such a 3D space would be highly adaptive to the movements, body significations, and the signalled needs of participants in the environment, and the AI supported space would account for complex environmental variables.

(iii) **Wide (creativity)-Narrow (AI) (WN)**

This intersectional space includes a wider and more expansive view of creativity inclusive of the place of creativity in the lives of learners, their imagination, their work with the collective and embodied practices that support creativity, as well as risk-taking. AI is narrow and intended to be supplementary to this expansive creativity, linking to risk-taking, imaginative work, group collaborative creative processes and recording of creative work. 'Intelligent' applications and algorithms could be part of the design to support the 'wide' creativity of learners.

Two examples of such possibilities in education might be:

- *Supporting the production, documenting, and recording of creative textual practices* such as creative writing in which language processing and virtual writing spaces might enable creative risk-taking, group collaborative writing practices and publishing spaces. AI in its task specific, narrow scope, can offer complementary ways of building, assessing, and editing text, which includes video materials, animation, and the multimodal presentation of text.
- *The creation and production of digital and interactive art* in which AI can enhance images and provide tools for exploring image variations and alternative possibilities for being visually creative (Mateas 2001; Mazzone and Elgammal 2019). AI can also, like textual production, provide a safe space for risk-taking and being iterative and productive from instances of failure. There is also the possibility in AI for the forms of presentation of digital art in virtual spaces.

(iv) **Wide (creativity)-Wide (AI) (WW)**

This interactional space is perhaps the most speculative and becomes an open ground in which expansive human creativity in all its aspects comes into (inter) play with the general (wide) notion of AI (Goertzel and Pennachin 2007). In this space the potential for AI to be agentially creative is taken seriously such that human and nonhuman creativities intersect and produce perhaps unpredictable outcomes.

Two examples of this space in education might be:

- *Unpredictable and unbounded creative outputs in which AI and human creativity are in synchrony* and form in the intersections and collisions between artificial creativity and human creativity. This might be in artistic and performative productions such as seen in theatre, writing and interactive virtual outputs.
- *Fully immersive and interactive AR and VR* teaching and learning environments in which the boundaries between the AI instructor/teacher/facilitator and the human are blurred. This might include immersive and realistic virtual excursions to art galleries, geographical sites and places of educational interest.

3.5.1 Concerns That Emerge from the Model

There is a note of caution attached to this thinking model. Currently, there is strong dispute as to whether there is such a thing as artificial general intelligence, with some philosophers and researchers suggesting that such a phenomenon is a long way off due to the significant problem of replicating human intelligence and even understanding the requisite characteristics of human intelligence that might be included in AI (Fjelland 2020). As such, the wide AI notion in this model can be seen as highly speculative—but important to think about, nevertheless.

When it comes to creativity, the potential of a general AI creativity or creative AI is in even more dispute, with compelling arguments about what it means for computers/machines/algorithms to have expansive (wide) creativity. Still, it is the role of a thinking model to explore this possibility and conceive of futures that might well evolve faster than might be presently thought.

Another concern is about the impact on human creativity of embracing AI. There is the possibility of AI, and especially how it is designed and implemented in teaching and learning, suppressing, or even stifling human creativity. This brings thinking about the intersections of creativity and AI back to the quality and appropriateness of design. It might also be the case that AI provides an avenue for crime and misappropriation of information that has impact in the educational sector. (Borenstein and Howard 2021).

There is also the issue of who designs the algorithms that impact education, in what might be seen as a potential bias that reflects the interests of global western corporations (Turner Lee et al. 2019). Ultimately, who controls design decisions? There is the pressing issue of the extent to which the emergence of AI might lead to the abrogation of responsibility in education and the giving of control to those beyond education.

3.6 Teaching and Learning Implications

Extensive application of AI is emerging in education, especially in the potential to be supportive and complementary to the learning of students and instrumental in the design and development of hybrid learning environments (Holstein et al. 2018). At the same time, the highly speculative rhetoric about what AI might do (in the wide or general sense) is not likely to be realised in education any time soon. The thinking model offered in this chapter highlights possibility-thinking in engaging with what AI might afford learners and their teachers in conceiving futures.

This possibility-thinking and critical examination of AI and its intersections with creativity and learning, should not be seen in a static way. Development in AI is incremental and moving, so that what is now deemed as 'narrow' may well include some 'wide' elements in the short to medium term. The sorts of AI that now operate in advertising, social media, security and identity-tracing, and language processing were not available a decade ago. While the problem of a self-aware and autonomous AI or machine (in the wide or general sense) seems a long way off, elements of autonomy in AI have emerged and will continue to emerge; therefore, it is important for educators to consider possibilities and applications, as well as the ethical, economic and practice issues that might surface.

To provoke further thinking about these possibilities I now offer three implications for schools, teaching, and learning that come out of the thinking model in Fig. 3.1. As mentioned previously, the examples given in this chapter are not the only ones, and I view these two implications as merely the start of a larger conversation about AI and its purposes in education.

3.6.1 Implication One: Human-Machine (AI) Interactions and Creativity

Clearly, the way that the digital, the virtual and the online have substantially impacted education is now a given. The recent impact of COVID-19 and the turning to online learning has reminded educators of the level of dependence on technologies and how teachers and learners intersect with such technologies. Sitting behind many of these technologies, and those emerging, is AI, often as a set of algorithms built on distributed computing networks. So, how can educators think proactively about how learners might be creative in juncture with the possibilities in AI? I believe that such junctures need to be designed so that the complementarity of AI can emerge in the creative potential of students in classrooms. This design might consider the narrow and wide conceptualisations suggested in the model, though with caution about the speculative concept of general AI.

3.6.2 Implication Two: Human-Machine Compatibility

Creativity, in its wide sense, is inclusive of the lives of learners and the creative expression that is meaningful to the learner. It is about belonging to the group and creating together as part of complex social and political systems. While AI researchers and advocates may well laud the possibilities in AI there is the problem of control and with this the ethics of AI in education (Holmes et al. 2021). Ultimately, the level of compatibility between AI and human creativity needs to be considered.

3.7 Conclusion

In this chapter I offer a thinking model for interrogating how AI might intersect in various spaces and dimensions with creativity, and how each might build on the other. The purpose is to be exploratory and to foster conversations in what is an emerging area of research and practice. I offer no definitive answers—merely my explorations, curiosities, and thorny questions. This thinking model is overtly oriented to education, teaching, and learning, but it might have more general application for contexts outside of education.

As part of introducing this thinking model, I surveyed the terrain of AI and visited some of the possibilities and issues that might transpire when considering this new area for critical thought and its junctures with learning and creativity. Not the least of these areas is considering what we mean by creativity, consciousness, and intelligence—difficult terrain indeed and pointing to the challenges (but exciting possibilities) in transacting the spaces and intersections between AI and creativity. I believe, despite its difficulty, that this terrain will have to be negotiated to move the debate forward and to make AI workable in practice.

The four spaces of intersection explicated as part of the model revolve around the notions of 'narrow' and 'wide' introduced to position various ways of seeing both creativity and AI. My delineation is, of course, arbitrary, but I believe it offers some possible ways into thinking about how creativity can be enhanced, supported, and envisioned considering the rapidly evolving prospects in AI and emerging futures in education.

References

Anderson, J. & Rainie, L. (2018). Artificial Intelligence and the Future of Humans. Pew Research Center. Retrieved from https://www.pewresearch.org/internet/2018/12/10/artificial-intelligence-and-the-future-of-humans/

Andrews, G. (2015). To Boldly Go Where No Learner Has Gone before: Independent Inquiry, Educational Technology, and Society. *Science Fiction. E-Learning and Digital Media*, 12(3–4), 343–360.

Bell, D. (2018). Ethics of Transference of Human Intelligence to Non-Biological Substrates. *Viterbi Conversations in Ethics*, 2(1). Retrieved from https://vce.usc.edu/volume-2-issue-1/ethics-of-transference-of-human-intelligence-to-non-biological-substrates/

Boden, M. (2018). *Artificial Intelligence. A Very Short Introduction*. Oxford University Press.

Borenstein, J., & Howard, A. (2021). Emerging challenges in AI and the need for AI ethics education. *AI Ethics* 1, 61–65. https://doi.org/10.1007/s43681-020-00002-7

Bown, O. (2012). Generative and adaptive creativity: A unified approach to creativity in nature, humans and machines. In J. McCormack, J. & M. d'Inverno, M. (Eds), *Computers and Creativity* (pp. 361–381). Springer.

Bozkurt, A., Karadeniz, A., Baneres, D., Guerrero-Roldán, A., & Rodríguez, M. (2021). Artificial Intelligence and Reflections from Educational Landscape: A Review of AI Studies in Half a Century. *Sustainability*. 13(2). https://doi.org/10.3390/su13020800

Brockman, J. (Ed.) (2020). *Possible Minds. 25 Ways of Looking at AI*. Penguin.

Bunz, M., & Braghieri, M. (2021). The AI doctor will see you now: assessing the framing of AI in news coverage. *AI & Society*. https://doi.org/10.1007/s00146-021-01145-9

Clarke, E. & Doffman, M. (Eds) (2017). *Distributed Creativity: Collaboration and Improvisation in Contemporary Music*. Oxford University press. https://doi.org/10.1093/oso/9780199355914.001.0001

Creely, E., Henriksen, D., & Henderson, M. (2021). Three Modes of Creativity. *Journal of Creative Behavior*, 5592), 306–318. https://doi.org/10.1002/jocb.452

Csikszentmihalyi, M. (1997). *The Psychology of Discovery and Invention*. Harper Perennial.

Csikszentmihalyi, M. (2014). *The Systems Model of Creativity*. Springer.

Dator, J. (2020). Editorial: Thoughts on Big Data, AI, and Futures Studies. *World Futures Review*, *12*(2), 151–155. https://doi.org/10.1177/1946756720913727

de Vasconcellos, S., da Silva Freitas, J., Junges, F. (2021). Digital Capabilities: Bridging the Gap Between Creativity and Performance. In S. Park, M., Gonzalez-Perez, & D. Floriani (Eds), *The Palgrave Handbook of Corporate Sustainability in the Digital Era*. Palgrave Macmillan, Cham. https://doi.org/10.1007/978-3-030-42412-1_21

Fjelland, R. (2020). Why general artificial intelligence will not be realized. *Humanities and Social Sciences Communications,* 7, 10. https://doi.org/10.1057/s41599-020-0494-4

Goertzel, B. & Pennachin, C. (2007). *Artificial General Intelligence*. Springer.

Hawkins, J., & Blakeslee, S. (2004). *On Intelligence*. Henry Holt and Company.

Holmes, W., Porayska-Pomsta, K., Holstein, K. *et al.* (2021). Ethics of AI in Education: Towards a Community-Wide Framework. *International Journal of Artificial Intelligence Education*. https://doi.org/10.1007/s40593-021-00239-1

Holstein K., McLaren B.M., Aleven V. (2018). Student Learning Benefits of a Mixed-Reality Teacher Awareness Tool in AI-Enhanced Classrooms. In C. Penstein Rosé et al. (Eds), *Artificial Intelligence in Education*. AIED 2018. *Lecture Notes in Computer Science*, 10947. Springer. https://doi.org/10.1007/978-3-319-93843-1_12

Höppner, G. (2021). Silhouettes analysis: a posthuman method for visualizing and examining the material world. *Qualitative Research*. https://doi.org/10.1177/1468794121999015

Horvatić, D., & Lipic, T. (2021). Human-Centric AI: The Symbiosis of Human and Artificial Intelligence. *Entropy*, *23*(3), 332. MDPI AG. https://doi.org/10.3390/e23030332

Hurson, T. (2007). *Think Better*. McGraw-Hill.

Kalyta, A., & Klymenyuk, O. (2018). Synergy of the flow of thinking. *Cognition, Communication, Discourse*, 15, 18–38. https://doi.org/10.26565/2218-2926-2017-15-02

Klein, E. (2021, March 30). Transcript: Ezra Klein Interviews Ted Chiang. *The New York Times*. Retrieved from https://www.nytimes.com/2021/03/30/podcasts/ezra-klein-podcast-ted-chiang-transcript.html

Köstler, L., & Ossewaarde, R. (2021). The making of AI society: AI futures frames in German political and media discourses. *AI & Society* (2021). https://doi.org/10.1007/s00146-021-01161-9

Kurzweil, R. (2005). *The Singularity is Near*. Viking Books.

Lee, S. & Trimi, S. (2021). Convergence innovation in the digital age and in the COVID-19 pandemic crisis. *Journal of Business Research*, 123, 14–22. https://doi.org/10.1016/j.jbusres.2020.09.041

Li, D., He, W. & Guo, Y. (2021). Why doesn't AI still have consciousness? *CAAI Transactions on Intelligence Technology*. https://doi.org/10.1049/cit2.12035

Mellamphy, N. (2020). Humans "in the Loop"? Human-Centrism, Posthumanism, and AI. *Nature and Culture*. https://doi.org/10.3167/nc.2020.160102

Mateas, M. (2001). Expressive AI: A Hybrid Art and Science Practice. *Leonardo*, 34 (2): 147–153. doi: https://doi.org/10.1162/002409401750184717

Mikalef, P. & Gupta, M. (2021). Artificial intelligence capability: Conceptualization, measurement calibration, and empirical study on its impact on organizational creativity and firm performance. *Information & Management*, 58(3). https://doi.org/10.1016/j.im.2021.103434.

Mitchell, M. (2019). *Artificial Intelligence. A Guide for Thinking Humans*. Penguin.

Mazzone, M., & Elgammal, A. (2019). Art, Creativity, and the Potential of Artificial Intelligence. *Arts*, *8*(1), 26. MDPI AG. https://doi.org/10.3390/arts8010026

Morasso, P. & Mohan, V. (2021). The body schema: neural simulation for covert and overt actions of embodied cognitive agents. *Current Opinion in Physiology*, 19, 219–225. https://doi.org/10.1016/j.cophys.2020.11.009

Moruzzi, C. (2021). Measuring creativity: an account of natural and artificial creativity. *European Journal forPhilosophy of Science,* 11(1). https://doi.org/10.1007/s13194-020-00313-w

Naudé, W. (2021) Artificial intelligence: neither Utopian nor apocalyptic impacts soon. *Economics of Innovation and New Technology*, 30(1), 1–23. http://dx. https://doi.org/10.1080/10438599.2020.1839173

Navidi, N., & Landry, R. (2021). New Approach in Human-AI Interaction by Reinforcement-Imitation Learning. *Applied Sciences*, *11*(7), 3068. MDPI AG. https://doi.org/10.3390/app11073068

Peeters, M., van Diggelen, J., van den Bosch, K., Bronkhorst, A., Neerincx, M., Schraagen, J. & Raaijmakers, S. (2021). Hybrid collective intelligence in a human–AI society. *AI & Society,* 36, 217–238. https://doi.org/10.1007/s00146-020-01005-y

Popenici, S. & Kerr, S., (2017). Exploring the Impact of Artificial Intelligence on Teaching and Learning in Higher Education. *Research and Practice in Technology Enhanced Learning*, 12(1). https://doi.org/10.1186/s41039-017-0062-8

Rust R., & Huang M. (2021). Artificial Creativity. *The Feeling Economy*. Palgrave Macmillan. https://doi.org/10.1007/978-3-030-52977-2_13

Sanders, N., & Wood, J. (2021). Combining Humans and Machines in an Emerging Form of Enterprise: the Humachine. *Foresight: The International Journal of Applied Forecasting*, 61, 28–35. https://search.ebscohost.com/login.aspx?direct=true&AuthType=shib&db=bth&AN=149583686&site=ehost-live&scope=site.

Sattin, D., Magnani, F. G., Bartesaghi, L., Caputo, M., Fittipaldo, A. V., Cacciatore, M., Picozzi, M., et al. (2021). Theoretical Models of Consciousness: A Scoping Review. *Brain Sciences*, *11*(5), 535. MDPI AG. Retrieved from https://doi.org/10.3390/brainsci11050535

Schlinger, H. D. (2003). The myth of intelligence. *The Psychological Record, 53*(1), 15–32.

Shear, J. (1997). *Explaining Consciousness: The Hard Problem*. MIT Press.

Soni, D. (2018, February 19). Introduction of evolutionary algorithms. *Towards data science. Retrieved from https://towardsdatascience.com/introduction-to-evolutionary-algorithms-a8594b484ac*

Staats, A., & Burns, G. L. (1981). Intelligence and child development: What intelligence is and how it is learned and functions. *Genetic Psychology Monographs,* 104(2), 237–301. https://psycnet.apa.org/record/1982-07433-001

Sternberg, R. (2020). *Human Intelligence: An Introduction*. Cambridge University Press.

Sternberg, R. J., Kaufman, J. C., & Grigorenko, E. L. (2008). *Applied intelligence.* Cambridge University Press. https://doi.org/10.1017/CBO9780511611445

Taulli T. (2021). *Future of AI. Implementing AI Systems.* Apress. https://doi.org/10.1007/978-1-4842-6385-3_9

Tegmark, M. (2017). *Life 3.0: being Human in the Age of Artificial Intelligence*. Vintage.

Turner Lee, N, Resnick, P. & Barton, G. (2019, May 22). Algorithmic bias detection and mitigation: Best practices and policies to reduce consumer harms. Brookings. Retrieved from https://www.brookings.edu/research/algorithmic-bias-detection-and-mitigation-best-practices-and-policies-to-reduce-consumer-harms/

Varghese M. (2021). *A Brief History of Creative Work and Plutonomy*. Springer. https://doi.org/10.1007/978-981-15-9263-8_13

Verma, A., Kumar, Y. & Kohli, R. (2021). Study of AI Techniques in Quality Educations: Challenges and Recent Progress. *SN Computing Science,* 2, 238. https://doi.org/10.1007/s42979-021-00635-3

Yi, G., & Loia, V. (2019). High-performance computing systems and applications for AI. *Journal of Supercomputing,* 75, 4248–4251. https://doi.org/10.1007/s11227-019-02937-z

Yolgörmez, C. (2021). Machinic Encounters: A Relational Approach to the Sociology of AI. In J. Roberge & M. Castelle (Eds), *The Cultural Life of Machine Learning*. Palgrave Macmillan. https://doi.org/10.1007/978-3-030-56286-1_5

Dr. Edwin Creely is a lecturer in the Faculty of Education at Monash University in Melbourne, Australia. His research focuses on creativity, critical literacy, digital literacies, and technology in learning. He has an emerging interest in online learning systems and the applications of artificial intelligence (AI) in education. Edwin is especially interested in innovation and creative practices and bringing new models and perspectives to educational research. He is widely published in creativity research and in education literature.

Chapter 4
East Asian Creative Ecologies in Networked Educational Worlds

Dan X. Harris

Abstract This chapter explores the ways in which establishing and maintaining creative ecologies requires a synthesising of 'online' and 'offline' practices, across diverse cultural, regional human and cultural lifeworlds. Braiding posthuman and creative ecology theoretics, this chapter argues that sustainable teaching, learning and living in speculative cultures requires whole-environment change, as well as mindset shift in restructuring the notion of lifespan education and creative labour. Drawing on a current study of East Asian and Australian creative education 'cultural creative clusters' and practices, this chapter advances a notion of creative ecologies which is not only transferable, but also specific to the differences of diverse time-space-practices.

Keywords Creative ecologies · Hong Kong · Singapore · Posthuman creativities · Creative and cultural industries

Educational learning—or the personal, political and social change made possible by critical education—is, as bell hooks reminds us, "always a vocation rooted in hopefulness. As teachers, we believe that learning is possible" (2013, p. xiv). This hope in the power of learning is nowhere more palpable than for those of us from minoritarian communities, or 'subaltern' geopolitical sites, like those featured later in this chapter. The work of.

> . . . truly decolonising minds and relationships regarding the function and possibilities of families (including families of education), we re-approach (as queer itself must continually do) the little publics of school, the education sector and families as 'learning communities', returning again to a utopian dream of what families might still become (Harris et al. 2017, p. 24).

As a queer American scholar, artist and activist living and working in the global south, I am aware of the privilege I hold in disrupting hegemonic discourses of

D. X. Harris (✉)
RMIT University, Melbourne, Australia
e-mail: daniel.harris@rmit.edu.au

D. Henriksen, P. Mishra (eds.), *Creative Provocations: Speculations on the Future of Creativity, Technology & Learning*, Creativity Theory and Action in Education 7,
https://doi.org/10.1007/978-3-031-14549-0_4

educational power, standardisation and creativity, and my imperative to do so. To that end, creativity studies has only begun to explore its true potential for expanding established practices and structures of global education.

Given the enduringly diverse definitions of creativity, I also pause to offer my own understanding of creativity, creative ecology, and agency here in order to provide a clear foundation for the discussion which follows in this chapter. Moving beyond individualist, human-centred and exceptionalist approaches to growing, understanding, and describing creativity, in recent years I have taken a creative ecologies approach to creativity studies. Even the concept of creative ecologies has begun to be taken up by a wide range of disciplines, including environmental education, actor network theory, digital media studies, and more. I define the ecologies as including human and non-human actors, always working together as an emergent "event, forever changing all of its elements as it co-creates the next moment" (Harris 2021, p. 5). This too is how I define creative agency, where all entities have their own agency, informing and co-creating in, with and through others. This heuristic is helpful, in particular, in highlighting the ways in which the preoccupation within creative education toward measuring creativity has not been successful in fostering creative thinking or practices (Harris 2021, p. 66).

Technology, too, remains a productively contested term, and so I pause to define my own understanding for the purposes of this chapter. I draw from Lev Manovich's characterisation of technology as "translating something into another format" (2001, p. 47). While the focus here is primarily on recent digital technology innovations and impacts on and within formal and informal education contexts, I remind readers of this broader understanding of technology in order to highlight that "changes in language and meaning affected through the 'logics' of computer technology in turn influence, shape and change the 'logics' of culture" (Holman Jones and Harris 2019, p. 73). As such, technology and culture are always already co-creating, and with these definitions in mind, I turn to a critical examination of the expectations and limitations of a utopian technological change in education.

It is no longer sufficient to look to digital or technological innovations for holistic 'solutions' to twenty-first century learning. For the last decade and more, educators have been seeking to define, understand, and foster creativity in education contexts, a task that has failed to move beyond twentieth-century individualist, humanist and exceptionalist approaches. Twenty-first century education requires a disruption of the old 'gifted and talented' model of creative education, and demands a more networked, collectivist restructuring of formal education and the ways in which creativity can help to get there. What Simonton has called the relationship between the *individual* and the *field* (2013), I have advanced as a model of *creative ecologies*, in which educators can make the conditions for creativity, through attention to the whole ecosystem – or field – of the learning environment, community, and collective practices. Twenty-first century creative education requires going beyond outdated notions of 'creative teaching' or 'creative learning,' and recognising it as a series of practices and mindsets that can be enhanced through a network or field.

As I have argued elsewhere, the failure of the education sector globally can be understood as.

> a communication breakdown between the micro and the macro work of creativity education, [and] also a failure to take a sufficiently systemic (or ecological) approach that considers school site, the school system and national education-and-economy strategies overall as a whole ecosystem that requires change, not just in the micro contexts of defining, assessing, or teaching a skill (Harris 2018, p. 68).

Claxton et al. (2008) have argued that the position of humans always already sits within more expansive 'megasystems' in which humans are inextricable and interdependent, and which humans and human innovation and creativity are co-creating with and within. But the move from a solely human-centred concept of creative capacitating, to an approach which recognises what Karen Barad has called agency, promises sustainable and more holistic progress. Posthumanism, or a theory of the more-than-human, is compatible with Barad's notion of agency as "not something that someone or something *has* to varying degrees, since I am trying to displace the very notion of independently existing individuals" (Barad 2012, p. 55), but rather a kind of *creative agency* (Harris 2021) that makes room for diverse creative affects, relationalities and practices.

This kind of collectivist approach to the agency inherent in 'un-commodified' creativity (Harris 2014), is at odds with anachronistic global education cultures that remain attached to the individual, to notions of individual success and to individualist trajectories, despite evidence attesting to the ways in which "individual endeavour is inextricable from the collective, and increasingly from algorithmic and digital networks" (Harris 2018, p. 71).

4.1 Posthuman School Networks

Many contemporary creativity studies scholars advance a range of theories and heuristics for an ecology of 'online' and 'offline' practices, across human and posthuman spheres, and across the wide range of sectors with which today's secondary students will no doubt find themselves engaged. But there is still a considerable gap between 'creative workforces' rhetoric, and networked school cultures (meaning both digitally-networked and socially-networked). How might empirical research help to bridge this gap?

In my 2014–2016 study on creativity in secondary schools, students overwhelmingly identified mathematics and computer teachers as the most creative in their school (Harris 2016). When questioned on what makes these classes or teachers the most creative, they noted the room for original thought, innovation, and the opportunities for bringing their 'whole selves' into those classroom ecologies. As an 'arts educator', I have rigourously reflected on my own practice and those of my arts education colleagues in wondering why arts students (not always but) often no longer feel this way about their arts classes. One reason may be the growing influence (and polarisation) of STEM subjects 'versus' arts subjects; in schools that are resource-poor, the Science, Technology and Mathematics classes are often resourced more generously than other subjects. Additionally, there are increasing external programs for enhancing and celebrating STEM achievement. A pervasive definition of creativity as something that is not only original, but having 'value' may be influencing educators' perspectives too.

Perhaps in classes like computing/robotics and STEM, students (and their teachers) are given more freedom and confidence because these areas are considered to have more 'value' under current national and global economic agendas; that is, because of their relationship to global workforces. It may seem obvious, but it remains a somewhat taboo topic in education and ITE (initial teacher education) conversations which insist that we do more in education than training workforces, as though training workforces is unsavoury. I still contend we do more, or at least have the potential of doing more. If educators could go further in consideration of networked creative education as the emergent, the encounter, or what Anna Craft called the 'trusteeship of ideas' (Craft et al. 2007, p. 28), both the processes of creative education, and the structures of networked learning, can be implicated across all learning domains and stages. I continue to wonder why educators and creative education scholars don't see the irony in attempts to standardise creativity (in practice, definition or assessment), given that the very nature of creativity rests with novelty.

My creative ecologies model goes beyond some other collective or networked theories about self/other in context, for example, actor-network theory; creative ecologies moves toward a more emergent conception of interconnectedness, and for this reason I draw on notions such as Karen Barad's *intra-action* ("the mutual constitution of entangled agencies", 2007, p. 33), which attend better to emergence. In actor-network theory, by contrast, both humans and their others pre-exist the encounter in which the creative event occurs (Cresswell et al. 2010). Barad describes agential realism as a kind of agency which "encompasses all the possibilities for reconfiguring the materiality of these and other forms" (in Harris and Holman Jones 2020, p. 22). Agency or agential realism urges perceptions of an event as the encounter between pre-existing entities, and rather as an emergent event (or intra-action) which has no past and no future but only exists in the moment, a notion that shares much with what some performance and creativity scholars have talked about as improvisation (Sawyer 2011) or flow (Csikszentmihalyi 1999).

4.2 A Posthuman Ethics of Creative Education

A posthuman approach to creative education understands that the creative ecology of a school includes not just human actants but also materials, spaces, buildings, practices, weather, atmospheres, and more. Kerry Chappell argues that "including and recognising other living beings, objects and environments too as embodied and therefore as enmeshed within the creative process better acknowledges the range of sources of action in creativity" (2018, p. 286). She rightly points out that a posthuman expansion of creative education requires "taking a new approach to what and who are seen as 'players' within the creative process, and to how the ethics of creative actions are understood" (p. 286).

Adopting a posthuman lens in considering creative education requires not just anthropomorphising the non-human elements in the creative ecology, a kind of projection of humanist concerns onto non-human others; rather, it demands a reconceptualization of the creativity rhetoric of 'use-value' when considering what and who is creative. It is not about attending to ecological or other kinds of sustainability in order to maintain a dominant humanist project, but rather an ecological creative education which attends to the interdependence of human and non-human collaborators and environments. Taken into current creative educational practices, this does not – like multiliteracies and collaboration – end at year 10 individual exams for assessing individual achievement. A true emergent ethics of creative learning ecologies would imply a structural and processual shift from individuals to collectives, not in order to banish or deny established notions of individual achievement and 'success' or 'failure', but to acknowledge the interdependent and collectivist nature of even that which appears to be individual achievement.

This is not, to my mind, antithetical to the project of 'workforce readiness' that education both pivots on and denies its alignment with. The irony is that this stubborn rhetoric hinges upon an idea that compulsory schooling is still a holistic and 'humanist' endeavour which is growing the whole person, the student who will learn social, ethical and critical ways of being in their school experiences which will prepare them for a more active citizenship once this phase is complete. No matter how good individual school or teacher practices are, however, the structures of – in particular – secondary schools continue to demonstrate a rigid adherence to siloed learning, individual endeavour, and increasingly over the last 10 years, a move away from criticality to a more neoliberal and indeed traditional orientation toward the '3 Rs' and STEM-focused learning. Strong and committed collaborative and multiliteracies work in the younger years falls away in years 11 and 12 when students are tested individually, predominantly still in writing, and in ways that reinforce neoliberal values.

4.3 Creative Ecologies in East Asia and Australia

Creative cities as a driver for modernisation and a mythical post-industrial future are a product of the 1990s global north (Gu et al. 2020). 'Creative industries', similarly, emerged from this shift from industrialisation to the knowledge economy in the late 1990's in the United Kingdom as a transactional means of explaining and identifying the 'soft' cultural practices that would subsequently drive the economy and replace production (for a good and more nuanced discussion of this emergence see the introduction to Gu et al. 2020). What is important for this chapter is to remember that these notions are all situated (as are all notions and conceptual frames) in particular times and places, and that not all times and places are the same. While questions of creative economies, production and post-industrial knowledge economic shifts have been circulating globally for over 20 years now, the globalising discourse of 'creative industries' implies that it is a conversation in which all in the global sphere are equally participating, which is untrue. This section is concerned with how the notion of creativity as an industry and economy is being considered and enacted in more recent times in the creative cities of Singapore, Hong Kong and Sydney, Australia, and how – importantly – education continues to struggle to be an agentic part of those conversations.

While the smaller South Asian countries have formed the economic coalition known as the ASEAN (Association of Southeast Asian Nations) coalition, there is much to be said about the digital (and social) media-driven networks of creative industry and ecologies with countries like Australia, which share a region but not strong cultural alignments. Nevertheless, the geographical (if not socio-political) proximities suggest important opportunities for mutual benefit in creative and cultural industries. In many ways, the countries of East Asia fall between China's superpower and expansionist agenda, and Australia and the attraction of the traditionally-dominant white western global north's creative cultural industries and the 'global' opportunities they promise. Kuan-Hsing Chen's call to decolonise cultural, industrial and theoretical thought-work in 'Asia' (itself a problematic term) challenges this region to untangle some of the onto-epistemological hangover from 500 years of colonisation from today's emerging Asian economies.

Creative and cultural industry scholars have been looking at exactly these issues, including Justin O'Connor who played a part in the original coining of the *creative economies* and *creative industries* discourse back in 1990s United Kingdom. *Creative Economies, Creative Cities* (2010, eds Lily Kong and Justin O'Connor) lays a range of diverse Asian-European perspectives on the differences and similarities between European, British and Asian approaches to notions and enactments of creative industries. More recently, O'Connor and Gu's *Red Creative: Culture and modernity in China* (2020) looks to city-sites like Shanghai as the vocabulary for understanding emerging creative cultures, as does my most recent study on creativity in East Asia and Australia.

Kong details the concept of 'cultural creative clusters' (2010) in Singapore, drawing on American and British discursive lineages, linking it to the (inaccurate) interchangeability of 'creative industries' and 'cultural industries' from those contexts. Following O'Connor & Kong (2010), Kong argues that creative clusters "succeed because of the development of tacit as opposed to formal, codified knowledge. As tacit knowledge is embedded locally, cultural producers thus need to be 'inside' the circuit of knowledge" (2010, p. 62). Kong goes further in unpacking the commodification of creative clusters from the arts and culture work itself, in her ethnographic work in Singapore. For Kong, 'cluster' in Singapore does not constitute physical proximity (what she calls 'aspatial'), but rather affinities; Kong's research examines "the ways in which clustering gathers meaning for visual artists. . .[and] how clustering is meaningful and adds value for the artists, whether it is in social and community terms, in artistic and creative value, or by way of economic benefit" (p 68).

As Chen reminds us, "The minjian tradition of networking and mutual support provides a basic mechanism for dealing with the harshness of life as a migrant worker in cities such as Beijing, Shanghai, and Guangzhou" (2010, p. 240). Chen – and O'Connor – look primarily to mainland China for their analyses, but my study immerses in Singapore and Hong Kong which (along with Taiwan) finds itself at a critical juncture with mainland Chinese expansion, and offers distinct approaches to building and educating for creative industrial futures.

4.4 Hong Kong

One high-level multinational tech company executive who was interviewed as part of this current study offered a nuanced portrait of both local as well as global ways of doing creative business in Hong Kong. She explains how, like the 'Microsoft model', this company also educates their own:

> The curriculum is – it's a completely free global curriculum. It's also the underlying curriculum for the Hong Kong Science and Technology Park, and they hit about seven and a half thousand students here. So, that's education and then, education goes from coding skills for kids in school, then into digital marketing skills for recent graduates into small businesses. So, we're going to run [program name removed for anonymity], which is training small businesses in how to use digital tools, even digital payments, because in Hong Kong, they don't get it, so we have to educate for that.

The digital marketing side includes topics such as:

> How do I get a [app name removed for anonymity] page? How do we use an ad? And then the third aspect for us is traditional marketers. We then look at educating traditional marketers in digital marketing skills. So, we run a thing called [app name removed for anonymity]. So, there's a [removed] program that people can sign up to, without their company sponsoring them, and do a six-month program, and then we sponsor a number of places, in (hangout) sessions as well, for a cohort of about 60 or so, and then that gets the message out as well, about what you can use (from our suite of products). And then we do also run a lot of global training. We are now split into two companies, it's [removed A] and [removed B]. [Removed A] is the other bits. It's where all the billions of dollars get spent on crazy things that [the CEOs] really wanted to do. The [removed B] part of the business which I'm in is where we make all the money so that they can spend it over there. And then, as well as that, we do lots of seminars for government. We'll run seminars for civil servants on how to use technologies. We'll run seminars for media, on how to use search tools to determine whether an image is real or fake. So, we run media labs, we educate legislators, because they're legislating technologies that they don't understand.

Increasingly as tech companies advise and train government, as well as petition and (as in Australia) block governments in order to allow their businesses to meet market demands, big business comes to exemplify how education is always a matter of 'educating *for*', rather than the still-pervasive myth of neutral, decontextualized, objective truths.

Back in the 1950s and earlier, when men (and sometimes women) worked for a single company for their whole working lives, and often were school-leavers trained by those companies, their skewed education was compensated for by the fact that they could, with some assurance, retain that job for their whole working lives, but the same is not true today. What in Hong Kong is known as 'slash culture' is increasingly a global phenomenon of working a range of different and mostly short-lived jobs in order to piece together a living and in some cases an independent creative life. But slash culture began as a twenty-first century economic imperative, not a lifestyle choice. It is characterised by a lack of economic certainty. And while formal education may be insufficiently educating for this kind of 'pick-up' global slash work culture, so is the corporate in-house training that claims to be doing education better. Possibly it's just a revisit of the mid-twentieth century in-house worker experience, without the employer loyalty that made those work-lives viable.

This tech multinational manager describes how her company maintains its creative and profitable ecology using a western model in an Eastern site:

> We provide support through partners, so we'll have channel partners that help them get on line, and we'll scale that out, basically, through partners. And then we also provide support for them in terms of creating community for those entrepreneurs as well. And then, the last one in the ecosystem, includes all the aspects that [company name removed] would be involved with. So cloud, hardware, ads, businesses and then our free products, our seven products that have a billion users each. So, we look at the overall health of the ecosystem, and one aspect of that ecosystem, which is creatively related is YouTube. So, we will run programs for Cantonese creators. So, if, for example, one of the team, our PR found this girl on Facebook who was doing some really interesting fitness videos, and encouraged her, instead of putting those videos just in Facebook, to put them on YouTube. She's now got 100,000 subscribers.
>
> In Hong Kong, YouTube would have 4.3 million active viewers, out of 7.3 million people, and they're watching over an hour a day. There's 65,000 creators making YouTube videos in Hong Kong on a regular basis. So, for us, the ecosystem is also about cultivating the creatives, and the creatives are the people who are producing really interesting stuff that can be consumed over the internet, and particularly video. And then the second aspect of cultivating the creators is mobile app developers, making sure they're doing really good interactive tools and creative agencies, that they know how to use our platforms to make marketing messages and storytelling more appropriate, and more engaging.
>
> Inside the organisation, developing our ecology can be anything from mentoring to coding to creativity to guitar playing, yoga. We actually set up our environment to try and encourage collaboration and creativity because we know that's where the good ideas are going to come from. So, for example the room that we're sitting outside of here, one night a week, it'll be a yoga studio and they just run free yoga classes for anyone that wants to come. We also embrace the concept of having downtime while in the office. So, we've got kitchens like this. We have micro kitchens within 50 paces of every desk so you can get snacks and drinks and when you're in those watercooler moments, you can have conversations that spark ideas and you bounce off each other. That's another thing that we do in terms of the physical environment.

4.5 Education for the Creative Future

Whether at the corporate (or educational institutional) ecology level, or the city-site level, as many creative and cultural industries scholars have noted, the infrastructure and investment in creative cities has turned out to be much longer-term than the original Creative Partnerships UK vision in the 1990s imagined. Education – including both the compulsory and higher education sectors – is one of these areas that has suffered from less consistent long-term investment, less innovation intent, and less structural reimagining than many other sectors (Craft 2011). Some continue to bifurcate the creative industry discourses and creative educational ones on the basis that multinational corporations are now seen as driving not only industry, but often curriculum change and education agendas more generally. AI developers, software companies, online network providers and other tech sector players are increasingly interested in providing their own education in-house, including through micro-credentialing which has rapidly expanded into higher education over the past few years and shows no sign of slowing (as in the example above). However, compulsory schooling and higher education degrees are still the current means of upskilling emerging workforces, even if they are delivering less of the critical and holistic education that the self-congratulatory 1970s through 1990s claimed it was committed to. So why aren't creative/cultural industries and creative education talking?

My study asks these questions in Singapore, Hong Kong and three sites across Australia. There is obviously a great deal of research on 'creative graduates' (that is, university graduates emerging from creative industries programs such as design, media, arts, and computer/games development), but what about others like 'non-digital' arts such as cultural festival curation, live performance, manual arts, and education, for example? In Hong Kong, several of our respondents agreed with one manager who noted that.

> Some of our Chinese colleagues maintain that, that philosophy of completing the syllabus results in wasting a lot of time. And they drill, drill, drill and it doesn't give the students anything new. And in fact, it would be more economical for them to take a more creative approach ...

the kind of approach increasingly seen as the workforce culture of the twenty-first century.

4.6 Singapore

Singapore, unlike Hong Kong, has been described by several participants as a paradox, in which.

> ... it definitely doesn't want to prepare people to be the technical people. We definitely do not want people to become the labourers, the cheap labour. We want them to be the thinkers, or the one who contributes to the thinking. But the environment is not providing or the policies or the way that we're brought up or the way that the system works doesn't allow us to move into those areas.

At each site of this study, 25 respondents were interviewed, a third of which were from higher education (in a range of disciplines). Of the higher education participants interviewed, 65% spoke of shifts within Singaporean education and society towards valuing arts and creativity more highly as a tool to enhance learning and developing inter- and intra-personal skills, noting especially that "*The larger community believes that art helps to make a more holistic self. We aren't quite there yet but we will be there.*" Two interviewees felt that these changes, while present, are minor and tokenistic. Both attributed this to a fear of artistic or creative freedom leading to potential criticism of government, which could in turn be damaging to schools:

> I think the absorption is slow because there's so much fear. There's fear that this is needed urgently but if really we're going to take it on for what it's worth, the changes have to be pretty radical. And it's political. It's hugely political. People don't want to be political and they don't want to take on the critical questions.

Thirty-five percent of Singaporean creative industry representatives interviewed commented on shifts within government and the education sector towards expanding the arts and creativity. One spoke of how more schools are becoming proactive in exposing their students to a wider variety of artistic endeavours including theatre performances and museums, but noted how it remains ad hoc and dependent upon individual schools and teachers, not sufficient change at the policy level.

Sixty-five percent of higher education representatives stated that rigidities of what one participant referred to as 'Singaporean efficiency culture' resulted in a stifling of creativity in the arts and education. *"To be product-driven is fine in itself. It is pressure I think, that this outlook plays out: we want it quick, want it soon, want it good. But we can't have all three."* One went on to question the government's motives for emphasising the creation of a vibrant city through the arts: *"Are we creating arts festivals and events in Singapore that contribute to the nation's arts and culture? Or are these merely creating noise? Noise in the sense that we're doing so many things, in the attempt to make Singapore a vibrant cultural city, but underneath this vibrancy, it is all noise."* Two articulated that rigid curriculum limits teachers' ability to teach in creative ways or give their students meaningful creative education experiences: *"I feel that there's really no point in them trying to create very exciting lessons, for example, which cannot be implemented in the classroom for various reasons."* It is important to note though that at this writing, Singapore has announced its first University of the Arts, an amalgamation of the famous LaSalle Institute and NAFA (Nanyang Academy of Fine Arts) Institute.

Fifty percent of the creative industries participants in Singapore spoke of the impact of neoliberal ideals on creativity in both education and industry. Three referred specifically to the education system as playing a role in limiting creativity, with a strong emphasis on outcomes and assessment, producing students who are answers-focused and unwilling to think individually and critically:

> . . . even when in the schools we try to infuse creativity into the curriculum interestingly you come up against student expectations because they are kind of thinking about 'okay what grade am I going to get and how are you going to mark me and am I going to get an A or C.' And sometimes I say it's the process, not the outcome, but students don't always get it.

One went further to state that it was not just students who struggled to think outside of the confines of assessment, but parents and teachers who pushed back against changes to assessment practices: *"you have teachers telling parents, "Don't panic. Don't panic. I'm going to set tests. I know that's what they're saying out there, but I'm still going to go for testing." So we have students in class going, "Why do I have to sit through ten weeks of this? Can you just give me some notes, and can we just take a test? Because therefore I only have to pass the test." And it's also about time: "Therefore I only need to spend 30 minutes, or an hour. Why do I have to go through the process?"*

4.7 Do We Still Think Technology Means Revolution for Education?

It becomes clear quickly when talking to a range of sectors in multiple sites that technology alone is not going to radicalise education, not even through the micro-credentialing and MOOC craze. Some reasons for this are obvious, but include clear cultural implications: technologies may be a transferable tool, but they are not used in the same ways in all places and times.

For one Hong Kong interviewee, a professional Cantonese opera singer from mainland China, social networks are more important than social media ones:

> Our Cantonese opera circle is very small. So, we need to see each other all the time. It's very important to cooperate, and it's important to keep a good social network. To know people. There is a Chinese saying, to know people is better to know the books, to read books. So, the social network is much more important than your knowledge.

For another who curates cultural festivals and teaches in this area in a Hong Kong university, the idea of diversity is crucial.

> Yeah, a shared culture is important, sure, but so are different areas of expertise in different areas that can all contribute. Like, how (mainland China's) Shenzhen being such an IT-driven place, and how would it have an impact on the performing arts, in developing the performing arts, and how Hong Kong can actually play a part in it? How can it play a part in dissemination through the use of technology of the performing arts to the rest of China for instance. There are lots of opportunities for kind of new joint cultural development.

There are important lessons to be taken from the proliferation of and opportunities afforded by creative-relational, digital networks and social media practices, individually, collectively and for both educative and marketing purposes. The organic nature of online network expansion, success, and the phenomenon of influencers, for example, are all evolving opportunities for expanding traditional education modalities. These phenomena usually happen – as we heard from the Hong Kong tech company executive – both through organic talents and content production, as well as through corporate sponsorship, development and backing. What lessons might education learn from these examples? Some combination of both investment (the 'innovation' part of creativity and innovation discourses), and relationality are necessary here. One medium-sized Hong Kong arts hub offers the following:

> How else do these networks form? I guess they also form organically through the programs that we organise. So particularly in the last two years, we did like some conferences and also some like workshops, and we kind of feed them in our masterclasses. So these were all held by practitioners who have a kind of methodology that they would like to share through workshops. So that also, in some very small ways, became the networks. We hosted this German artist, a wonderful lady who brought in a researcher who was like a co-host for this workshop. And how it evolved was because I'd met them in a workshop in Kuala Lumpur, which was their network. And then we did one here, and then she did one in Penang. So then, in a way, the networks grew – flowed into each other. And because of that, the participants are co-hosts of those projects then became collaborators at a later stage. So I found it quite interesting over time to see how that flow of energy sort of worked organically.

How often do we ask such questions in the planning of education, teaching and learning? Are there more opportunities for asking how to grow the networks (both digital and non-digital) of our students? A creative ecological approach which includes digital networks and infrastructure is just one part of larger metasystems that promise to bring compulsory education up to speed for training both more critical global citizens, as well as more effective twenty-first century workforces.

4.8 Conclusion

The notions of cultural creative clusters, and creative ecologies, can be applied to learning environments and practices to great effect. Contemporary education offers many urgent and long term possibilities for creative and networked innovation, opportunities that go far beyond material and technological advances. By capitalising on lessons learned from COVID-19, the creative turn in industry and economics, and trends in micro-credentialing, a creative ecologies approach to fostering creative environments takes a more networked and joined-up approach to nurturing both creative workforces and critical citizenship.

It seems that if education remains recalcitrant toward creative change and the need for digital, collaborative working styles, the so-called 'real world' will continue to move on without us. Global workforces are rapidly moving away from traditional markers of educational achievement, including full course degrees. From corporate self-credentialing, in-house training and 'untrained' start-up needs, to 'slash culture' multi-sector jobbing, to micro-credentialing like badges and MOOCs, twenty-first century workforces are accelerating away from the business-as-usual of standardised testing and old-school single-disciplines, and toward more agile, more multi-disciplinary and more industry-linked creative ecological lives.

Acknowledgement This research was funded by the Australian Government through the Australian Research Council Future Fellowship scheme, project #FT170100022 (2017–2021).

References

Barad, K. (2012). Interview. In R. Dolphijn & I. van der Tuin (eds.) *New materialism: Interviews and cartographies.* Open Humanities Press, pp. 48–70.

Chappell, K. (2018). From wise humanising creativity to (posthumanising) creativity. In (Eds.) Snepvangers, K., Thomson, P. and Harris, A., *Creativity Policy, Partnerships and Practice in Education.* Palgrave Macmillan, pp. 279–306.

Chen, K. (2010). *Asia as Method: Toward Deimperialization.* Duke Univ Press.

Claxton, G., Craft A., & Gardner, H. (2008). Concluding thoughts: Good thinking-education for wise creativity. In A Craft, H. Gardner, G. Claxton (Eds), *Creativity, wisdom and trusteeship: Exploring the role of education.* Corwin Press, pp. 168–176.

Craft, A. (2011). *Creativity and education futures: Learning in a digital age.* Trentham.

Craft, A., Gardner, H., & Claxton, G. (2007). *Creativity, wisdom and trusteeship: Exploring the role of education.* Sage.

Cresswell, K.M., Worth, A. & Sheikh, A. (2010). 'Actor-Network Theory and its role in understanding the implementation of information technology developments in healthcare.' *BMC Med Inform Decis Mak* **10**, 67 (2010). https://doi.org/10.1186/1472-6947-10-67.

Csikszentmihalyi, M. (1999). '16 implications of a systems perspective for the study of creativity.' (Ed. Robert Sternberg) *Handbook of creativity*, Cambridge Univ Press, pp. 313–335.

Gu, X., Lim, M. K. & O'Connor, J. (2020). *Re-imagining creative cities in twenty-first century Asia.* Palgrave.

Harris, D. (2021). *Creative Agency.* Palgrave.

Harris, A. (2018). 'Creative Agency/creative ecologies'. In (Eds.) Snepvangers, K., Thomson, P. and Harris, A., *Creativity Policy, Partnerships and Practice in Education.* Palgrave Macmillan, pp 65–87.

Harris, A., Holman Jones, S., Faulkner, S., and Brook, E. (2017). *Queering Families/Schooling Publics: Keywords.* Routledge.

Harris, Anne. (2016). *Creativity and Education.* Palgrave Macmillan.

Harris, A. (2014). *The creative turn: Toward a new aesthetic imaginary.* Sense.

Harris, A. & Holman Jones, S. (2020). *The queer life of things: Performance, affect, and the more-than-human.* Routledge.

Holman Jones, S. & Harris, A. (2019). *Queering Autoethnography.* Routledge.

hooks, b. (2013). *Teaching community: A pedagogy of hope.* Routledge.

O'Connor, J. and Gu, X. (2020). *Red creative: Culture and modernity in China.* Intellect.

O'Connor, J. & Kong, L. (2010). *Creative Economies, Creative Cities.* Palgrave/Springer.

Manovich, L. (2001). *The language of new media.* MIT Press.

Sawyer, R. K. (2011). *Explaining creativity: The science of human innovation.* Oxford University Press.

Simonton, D.K. (2013). What is a creative idea? Little-c versus big-c creativity. In K. Thomas & J. Chan (Eds.) *Handbook of research on creativity.* Edward Elgar Publishing, pp. 69–83.

Dan X. Harris is Professor, Associate Dean, and Australian Research Council Future Fellow at RMIT University in Melbourne, Australia. Their research focuses on the intersection of creativity at both practice and policy levels; on cultural, sexual and gender diversities; and on performance and activism. They are committed to the power of collaborative creative practice and social justice research to inform social change. Harris is most widely known for their scholarship in creativity studies, affect theory and autoethnography and has authored over 100 articles/book chapters, 17 books, plays, films and spoken word performances.

Chapter 5
Room to Run: Using Technology to Move Creativity into the Classroom

Jonathan A. Plucker, Melanie S. Meyer, Sareh Karami, and Mehdi Ghahremani

Abstract In a rapidly changing world, it can be hard to know where to invest resources in K-12 education to prepare students for the future. Educators are pulled in many different directions as they try to address course content, local curriculum, state standards, and accountability testing in a very limited amount of instructional time. Integrating creativity, innovation, and entrepreneurship (CIE) education into the K-12 school day is a promising approach that can give students from all cultural, linguistic, and economic backgrounds access to advanced learning opportunities that are enhanced by technology. Yet many educators believe that any focus on CIE distracts from the standards- and test-focused criteria against which their effectiveness is judged. This chapter describes how the strategic use of technology provides teachers with room to run as they apply principles of CIE education across the curriculum, providing students with opportunities to develop creative process skills, innovate within areas of interest, and prepare for the challenges of rapidly changing postsecondary education and career paths. Combining technology and CIE learning experiences can help K-12 educators pursue both equity and excellence.

Keywords Creativity · Innovation · Entrepreneurship · Equity · Excellence gaps · Integrated technology

J. A. Plucker (✉) · M. S. Meyer
School of Education and Center for Talented Youth, Johns Hopkins University, Baltimore, MD, USA
e-mail: jplucke1@jhu.edu

S. Karami · M. Ghahremani
Department of Counseling, Educational Psychology, and Foundations, Mississippi State University, Starkville, MS, USA

D. Henriksen, P. Mishra (eds.), *Creative Provocations: Speculations on the Future of Creativity, Technology & Learning*, Creativity Theory and Action in Education 7,
https://doi.org/10.1007/978-3-031-14549-0_5

Classroom teachers in K-12 schools have the task of not only preparing students for the next grade level but also ensuring that they have the competencies required for postsecondary education and career paths. This responsibility is made even more challenging by a number of constraints, including mandated curricula and standardized testing (Beghetto 2013; Omdahl and Graefe 2016). Many educators believe that they must focus on content delivery and test preparation in order to earn favorable evaluations as part of school accountability systems and to ensure student readiness. Teachers do have a limited number of instructional days each school year, so when it comes to deciding between reinforcing required content and bringing creativity into the classroom, more often than not, creativity falls by the wayside (Beghetto 2013). Contrary to popular belief, teachers do not have to choose between creativity and content. In fact, bringing creativity into the classroom through integrated technology can give teachers the freedom to embed course content into student-centered instructional approaches that build problem-solving skills and prepare students for the future.

The world is changing faster than ever before and with that change come social, economic, technological, and environmental challenges ("Sustainable Development Goals" 2017). Some high-demand occupations (e.g., social media managers, data miners, cloud architects) did not exist a decade ago and many school-aged children will likely work in career fields that are still emerging (Global Business Coalition for Education [GBCE] 2018); World Economic Forum [WEF] 2020). This rapid change will require the next generation to invent ways to navigate an uncertain future. Although this uncertainty could cause apprehension, in many instances it has been a catalyst for positive changes in public policy related to education. Supporting creativity and innovation has become a priority in global discussions (e.g., "P21 framework definitions" [P21] 2009; WEF 2020) and a range of projects have been initiated to enhance student access to creative and innovative learning experiences (e.g., "Europe 2020 flagship initiative: Innovation union" 2011; United Nations Educational, Scientific, and Cultural Organization [UNESCO] 2015).

Universities increasingly recognize that their graduates need to (a) understand social challenges, (b) identify opportunities to address them, (c) develop new or improved products and services for addressing those challenges, and (d) competently advocate for those ideas (Grossman et al. 2011; "Sustainable Development Goals" 2017). Individuals who can create and apply new knowledge in novel ways will have an economic advantage over those who cannot (Powell and Snellman 2004), so educating a creative, innovative, and entrepreneurial workforce has become a new direction for twenty-first-century universities around the world and one that is backed by policymakers, global organizations, and employers (National Association of Colleges and Employers [NACE] 2018). A "renaissance of global creativity in education" (Guo and Woulfin 2016, p. 158) can ensure that elementary and secondary students have the necessary skills and mindsets to prosper in higher education and the evolving labor market.

Just as universities are adapting to equip young adults for the demands of the workforce, K-12 schools can also prepare children for postsecondary education and careers through creativity, innovation, and entrepreneurship (CIE) education.

Despite a clear need for elementary and secondary students to develop creativity and problem-solving skills prior to high school graduation, educators perceive several barriers to integrating CIE and inquiry-based learning into existing curricula. High-stakes test preparation, locally determined instructional priorities, limited opportunities for relevant professional learning, and time constraints can all make it difficult for teachers to implement new pedagogy to foster creativity (Beghetto 2013; Lee et al. 2021; Omdahl and Graefe 2016), but these challenges are not insurmountable.

5.1 Creativity, Innovation, and Entrepreneurship (CIE) in the Classroom

Strategic use of technology can give teachers room to run as they apply principles of CIE education across the curriculum. As educators present students with opportunities to develop creative process skills and innovate within areas of interest, they are also preparing them for the challenges of rapidly changing postsecondary education and career paths. Creativity, innovation, and entrepreneurship are similar in that they each involve the creation of novel and useful solutions, but there are distinct differences between the constructs (Dino 2016; Shalley et al. 2015).

5.1.1 Creativity

Creativity is often described as "the interaction among aptitude, process, and environment by which an individual or group produces a perceptible product that is both novel and useful as defined within a social context" (Plucker et al. 2004, p. 90). In other words, creativity happens when personal factors (e.g., motivation, persistence) interact with contextual factors (e.g., organizational flexibility, risk tolerance) to produce original ideas that solve a problem or fill a need (Shalley et al. 2015). Classroom instruction that features creative process skills (e.g., problem finding, knowledge acquisition, information gathering, incubation, idea generation, combining ideas, selecting the best ideas, externalizing ideas; Sawyer 2012) can foster engagement as students pursue topics of interest. In addition, working through the recursive stages of the creative process allows teachers to embed content-specific knowledge and skills to address state standards within inquiry-based lessons (Lee et al. 2021). Technology can facilitate information gathering, but digital tools (e.g., social media, document sharing, learning management systems) can also give students and teachers virtual collaboration platforms to share early-stage ideas and manage feedback as part of the creative process and the classroom assessment process.

5.1.2 Innovation

Creativity involves "the successful generation" of a product that is considered original and useful in a specific context, but innovation goes a step further and includes "the successful implementation of such an idea" (Dino 2016, p. 26). The Stanford Design Thinking Model ("An introduction to design thinking," 2010), which has been applied in professional, postsecondary, and K-12 settings is composed of several non-linear stages that include empathizing with the end-user, defining parameters, generating ideas, prototyping, testing, and assessing. Innovation extends creativity by adding several steps, including troubleshooting and revising prototype designs. The design process provides a framework that educators can use to scaffold individual student learning and deliver mandated course content through hands-on "learning through making" (Giannakos et al. 2017, p. 78) experiences, rather than just traditional lectures. Bringing the design process into schools and classrooms through maker spaces and inquiry lessons can provide educators with the flexibility to offer "low floor" (e.g., accessible to all ability levels) and "high-ceiling" (e.g., increasingly complex projects over time; Giannakos et al. 2017, p. 81) options to differentiate for individual learners within the classroom. Technology tools can support the design process by facilitating digital collaboration and information gathering, but they can also allow students to produce and test their designs (e.g., 3-D printing, digital simulations).

5.1.3 Entrepreneurship

Entrepreneurship extends the work of creativity and innovation by requiring individuals to (a) identify a need for the creative or innovative product in the economy, (b) locate potential markets in which to promote the goods or services, and (c) plan for the ups and downs of supply and demand in the market (Shalley et al. 2015; Venkataraman 1997). Moving creativity into the real world through entrepreneurship education is one way to extend learning experiences beyond the classroom. When educators allow students to find real problems in their schools and communities and provide a framework for developing and testing solutions with end-users, learning experiences are transformed and students have an increased sense of purpose (Omdahl and Graefe 2016). As students work through iterative processes for creating (e.g., ideation, articulation), implementing (e.g., prototyping, troubleshooting), and marketing their ideas (e.g., needs evaluation, user specifications), educators can strategically embed mini-lessons and project components that also address curriculum standards and skills tested on standardized assessments (Lee et al. 2021). Technology can facilitate these processes by enabling students to communicate with end-users, collaborate with team members and consumers on product design, and construct solutions.

5.1.4 Combining Creativity, Innovation, and Entrepreneurship in Education

Although there are differences between creativity, innovation, and entrepreneurship, these constructs may best be viewed as "subdomains in a larger value-adding ecosystem" (Dino 2016, p. 142). In other words, the individual components of CIE education all share similar process skills which, over time, can be taught, applied, refined, extended, and enhanced by technology, making CIE curriculum a flexible approach that can be implemented with students across all grades and ability levels.

For the purposes of this chapter, *creativity* refers to engaging in a process to *generate products and ideas* that are original and appropriate for the context (Plucker et al. 2004; Sawyer 2012). *Innovation* refers to engaging in the design process to *implement creative ideas* by developing and testing products or services (Dino 2016; "An introduction to design thinking" 2010). *Entrepreneurship* refers to the process of *analyzing market and end-user needs in order to effectively promote, distribute, and support* creative and innovative products or services (Dino 2016; Shalley et al. 2015; Venkataraman 1997). In school settings, technology-enhanced CIE learning experiences can provide a vehicle for educators to (a) embed mandated course content, (b) prepare students for required testing by reinforcing skills in applied contexts, and (c) help students develop non-cognitive competencies (e.g., psychosocial skills, executive functioning) so they have the skills to excel as school and work environments present greater challenges.

5.2 Integrating Technology into the Creative Process

In today's classroom, the roles of creativity and technology are intertwined in a reciprocal relationship. On one hand "technological change is driven by human creativity" and on the other hand, technology "provides new contexts and tools for creative output" (Henriksen et al. 2016, pp. 27–28). Technology has opened up an array of possibilities to improve teaching practice, including discovering, creating, and sharing ideas and content (Henriksen et al. 2016; Mishra et al. 2013). Integrating technology in schools goes beyond teaching students to use digital devices and navigate virtual platforms. The true value of integrating technology into education is in the ways it can facilitate domain-specific content learning, creativity, and connectedness. Educators must have technological, pedagogical, and content knowledge, but they also need to look for opportunities to "repurpose existing tools" (e.g., social media; Henriksen et al. 2016, p. 31) and provide learning experiences that allow students to use technology tools to enhance self-directed and collaborative design projects (Mishra et al. 2013).

The International Society for Technology in Education (ISTE) has developed seven standards for students, which include becoming (a) empowered learners, (b) digital citizens, (c) knowledge constructors, (d) innovative designers, (e) computational thinkers, (f) creative communicators, and (g) global collaborators ("ISTE standards for students" 2021). These standards align closely with many of the skills required for creativity, innovation, and entrepreneurship (e.g., flexible thinking, reserving judgment, evaluating criticism). When technology is available to learners during the school day and beyond, it can be used to enhance learning and the application of that learning by giving students digital spaces to create and innovate beyond the walls of the classroom (Henriksen et al. 2016; Lee et al. 2021; Mishra et al. 2013). Given that key outcomes of technology education and CIE education overlap (e.g., creative process skills, design process skills, communication skills), using technology to integrate creativity into the school day can give students from all cultural, linguistic, and economic backgrounds access to experiences that promote both creativity-related and technology-related skills.

5.3 Pursuing Excellence and Equity Through Technology-Enhanced CIE Learning

Excellence gaps are differences in advanced performance among students from different cultural, linguistic, and economic backgrounds. These gaps at the highest levels of achievement can be related to a number of factors, including differences in resources (e.g., financial, information, technology) and variable access to advanced learning opportunities (e.g., talent development programs, online learning platforms; Plucker et al. 2010, 2013). Creativity-focused instruction can benefit all students, but opportunities for CIE learning are often offered in out-of-school-time through extracurricular activities and special programs that may not be available to all students, particularly those with limited resources (e.g., transportation, program fees; e.g., Snellman et al. 2015; Weininger et al. 2015). Extracurricular programs can help students develop problem-solving skills that are useful for addressing personal and societal challenges, and there is some evidence that these programs can contribute to increased academic achievement (Omdahl and Graefe 2016; Snellman et al. 2015). However, because these options are not available to all students, they may also perpetuate excellence gaps (Covay and Carbonaro 2010; Snellman et al. 2015). Investing in CIE education by making it part of the K-12 school day, rather than an add-on extracurricular program, is one approach that can be used to address these challenges.

In addition, access to digital devices, Internet connections, and technology training can vary drastically from individual to individual and school to school. One-to-one initiatives that provide take-home devices for all students and teachers have been associated with achievement gains in core academic subjects (e.g., language arts, math, science, social studies) when that technology is integrated into a larger program of social and academic support in the learning community (Zheng et al. 2016), which suggests that just having access to digital devices is not enough. In two meta-analyses, technology-enhanced literacy instruction (Cheung and Slavin 2012) and mathematics instruction (Cheung and Slavin 2013) were more strongly related to academic growth than computer-delivered instruction alone, which emphasizes the fact that technology can be a tool to improve learning outcomes, but the mere presence of technology in the learning space does not guarantee improved educational outcomes. Taken together, this research highlights the importance of integrating technology into a larger system of support, such as CIE education during the school day. CIE provides an engaging curriculum, opportunities to create and innovate, and interaction with educators, peers, and end-users, all of which can be facilitated with a variety of technology tools. Since access to technology can vary widely across school districts and individual campuses, educators need to be mindful of this and other equity and access issues when planning CIE instruction with integrated technology.

5.4 Reducing Excellence Gaps

Schools can help close excellence gaps through a combination of research-based strategies at the policy-level (e.g., support for advanced learning in school accountability systems), the district and school-level (e.g., relevant professional learning, universal screening for advanced academic programs, use of local norms for placement, expanded advanced learning opportunities), and the classroom level (e.g., frontloading to prepare students for advanced learning, grouping to facilitate acceleration, and psychosocial interventions; Plucker et al. 2017; Plucker and Peters 2016). These strategies address equity and reduce excellence gaps by giving students with academic needs not met in the traditional classroom options for learning environments that do meets their needs. Technology-enhanced CIE education that occurs during the school day can go a long way toward promoting both equity and excellence. The self-directed nature of CIE projects allows individual students to pursue topics of interest, which is likely to increase engagement in learning course content. Educators can provide varying levels of guidance for students as they work through challenges in the problem-solving and design processes. These flexible inquiry- and problem-based learning experiences allow students with advanced learning needs to dive into increasingly complex projects that challenge their current skills and abilities, which can lead to academic growth, psychosocial strength, and an understanding of the rigors of engaging in related professional domains. So, not only does CIE education give teachers the ability to deliver required content and prepare students for mandated testing, but the framework also has the capacity to provide differentiated learning experiences for students with advanced learning needs and to allow them to begin making choices about talent domains they might pursue after high school.

5.5 Benefits of Technology-Enhanced CIE Curriculum

Although there may be obstacles to implementing CIE education across the curriculum during the school day, there are very real benefits to providing this type of instruction. CIE curriculum with integrated technology can support school accountability goals, promote equity and excellence, and prepare students for the future.

5.5.1 CIE Can Support School Accountabilityspiepr146 Goals

Standardized Testing The No Child Left Behind Act (NCLB 2001) required the use of standardized testing in school accountability systems, and since that time, educational testing has become a focus in American education (Koretz 2008). Standardized test scores are used to inform the allocation of resources and to hold schools and teachers accountable for providing a high-quality education to all students. However, implementing a CIE curriculum is unlikely to hurt test scores. In fact, the emphasis on problem-finding, problem-solving, and flexible thinking has the potential to increase student achievement (Omdahl and Graefe 2016). Although teachers may feel constrained by the need to prepare students for these tests, using technology to bring creativity into the classroom can allow instruction to move beyond teaching content for a test to applying that content to real-world problems in innovative ways that lead to deeper understanding (Mishra et al. 2013). For example, some science content, such as the water cycle, is taught and tested in multiple grade levels. Rather than rehashing the same steps and vocabulary, educators can create problem-based learning experiences that ask students to discover factors impacting local water sources, create solutions, and share their ideas with community stakeholders. Students will not only be prepared to answer questions about the water cycle on standardized tests, but they will also gain experience using technology tools to discover, create, and share solutions to improve water quality in their community.

College and Career Readiness CIE can also help schools meet accountability requirements related to college and career readiness (Hackmann et al. 2019). When opportunities for creativity and innovation are incorporated across the curriculum, educators can expose students to career paths in humanities, the arts, public service, and STEM domains. Videoconferencing technology allows educators to bring speakers into the classroom to discuss how they use domain-specific skills in their professions. For example, geometry teachers could invite an interior designer or a landscape architect to discuss how they calculate the amount of paint, flooring, or mulch needed for a design project. English teachers could invite an attorney or an advertising account manager to describe how they use argumentation strategies as part of their jobs. These guest speakers can not only reinforce real-world applications of the curriculum, but they can also help students imagine a future as a professional in the field. Educators can use technology-enhanced CIE during the school day as a springboard to help students identify areas of interest and explore compatible postsecondary education and career paths. In higher education and the workforce, students will need the ability to generate new ideas and turn them into solutions through creative problem-solving (NACE 2018). CIE learning experiences focused on discovering real-world problems, creating new applications for course content, and sharing their discoveries with peers and professionals can ensure that students have the confidence that they can excel in a variety of postsecondary environments.

5.5.2 CIE Can Promote Equity and Excellence

Course-Embedded Options If students have to opt-in to programs in schools, there are a variety of barriers to participation, including scheduling or a lack of awareness about the opportunities and why they might be beneficial. If CIE education is integrated across core curriculum areas (e.g., language arts, math, science, and social studies) in elementary and middle school, then all students would have exposure to creative process skills and the design process (Omdahl and Graefe 2016), as well as psychosocial skills coaching to learn to receive and incorporate feedback about their creative ideas (Rinn 2020; Subotnik et al. 2011). Due to graduation requirements, it can be harder for high schools to integrate CIE using a cross-curricular, whole-school approach. Service-learning programs, such as the College Board's Advanced Placement (AP) with WE and International Baccalaureate's (IB) Creativity, Activity, Service (CAS) units, can be embedded into individual courses. The AP with WE program provides online learning modules to scaffold the creation of community service projects related to course content ("AP with WE Service" 2021) and IB CAS service-learning projects are developed by individual schools and classrooms ("From the classroom to the community" 2015). For courses not under the AP or IB umbrella, educators can develop curriculum units to integrate CIE principles into existing content with technology tools to facilitate student research, idea development, and collaboration activities.

Equity-Building Partnerships Community partnerships are one way that technology-enhanced CIE can be supported. Businesses can provide resources (e.g., financial support, technology) and expertise (e.g., mentors) for schools as part of outreach and community engagement efforts. The private sector and schools can both benefit from such partnerships. Corporations can provide students with a chance to engage in the real challenges that companies face in the course of doing business, and businesses can help students develop an awareness of the knowledge and skills required to operate in the domain (Business Roundtable 2018; WEF 2020). When schools create more opportunities for students to learn through CIE and integrated technology during the school day, teachers need quality professional learning experiences and instructional support to provide meaningful and effective instruction (Henriksen et al. 2016). Training that emphasizes equity-driven pedagogy and empowers teachers to help students dive into real-world problems and create change at the classroom, campus, and community level can strengthen cross-curricular CIE implementation (Spoon et al. 2020; Novak et al. 2020). University partnerships may provide another option to support teacher professional learning for CIE education and technology integration.

5.5.3 CIE Can Prepare Students for the Future

Frontloading Students are more likely to be successful in advanced learning programs if they are prepared. Frontloading includes pre-teaching foundational content and skills to prepare students for future advanced learning opportunities (Plucker and Peters 2016). For example, there are certain mathematical concepts students need to master before they are placed in advanced or accelerated math courses. Frontloading is a proactive approach that can be used to start teaching those skills in elementary school so students are prepared for accelerated content in middle and high school. This gives them the ability to take the math courses (e.g., Algebra II, Calculus) required for access to many college STEM and business programs. CIE education in elementary and middle school allows students to engage in areas of interest while teachers use the creative process to front load content and prepare students for advanced-level coursework in high school and beyond. However, this approach requires coordinated efforts at the district and campus levels to ensure that policies, school schedules, curriculum, and resources are in place to support these efforts (Lockhart et al. 2021). Technology can be especially helpful in the implementation of frontloading because each student may need a distinct curriculum and personalized learning assistance. Adaptive technology helps teachers individualize frontloading efforts by providing platforms to differentiate curriculum, assessment, and growth-focused feedback for each student. Recent research suggests that such differentiation in the absence of technology is quite difficult for the vast majority of teachers (Callahan et al. 2020; Plucker and Callahan 2020).

Creative Articulation Beyond foundational content and skills, students also need the ability to advocate for their creative ideas and products. Creative articulation is the "process through which an individual influences the numerous elements that interact to determine the acceptance or rejection of a creative product" (Plucker 2016, p. 152). Creativity, innovation, and entrepreneurship are complex domains that require students to accept criticism and use it to improve their ideas or products. CIE programs can help students develop the psychological (e.g., motivation, self-beliefs) and social skills (e.g., giving, receiving, incorporating feedback) that individuals need to engage in challenging endeavors (Rinn 2020), such as creating, inventing, and marketing a product or service. Technology integrated into CIE instruction supports creative articulation, in large part, because it allows students to use platforms similar to those used by professional creators, innovators, and entrepreneurs to develop, refine, and promote their work. For example, it is anachronistic for a student with a good idea to share it by writing an early 1800s-style pamphlet. That would be silly, of course, but is it really any less appropriate than having a student share an idea in a five-paragraph essay when they could be using a suite of multimedia and networked technologies to share the idea? Using technology to help students learn how to advocate for their work and provide constructive criticism regarding the work of others is a major creativity-fostering skill.

5.6 Technology-Enhanced CIE Programs

There are many CIE initiatives worldwide, but two programs in the United States have a strong track record of supporting STEM talent development and reducing excellence gaps by expanding advanced learning opportunities through university and community partnerships, frontloading critical content and skills for success in STEM domains, and supporting teachers with professional learning and networking. These initiatives combine creativity, innovation, and entrepreneurship with communication, design, and production technology to provide engaging learning experiences during the school day. An advantage to the use of such programs is that school district leaders, who may not see the value in bringing creativity into education, may be more likely to appreciate technology-focused programs. CIE initiatives that emphasize creativity, integrate technology, and implement creative ideas allow students to develop both technology-related and creativity-related skills that policymakers and education stakeholders can all get behind.

5.6.1 Lemelson-MIT InvenTeams Program

Since 1994, the Massachusetts Institute of Technology (MIT) School of Engineering has partnered with the Lemelson Foundation to offer the InvenTeams grant program (https://lemelson.mit.edu/inventeams). Teams of 10 or more high school students and a faculty sponsor go through an initial application process. If selected, they have 5 months to develop and submit a final application detailing an invention that meets a need they identified in their community. Ten to 15 teams of finalists are awarded $10,000 grants and have approximately 8 months to build a working prototype of their invention to share at the InvenTeams showcase, EurekaFest. The InvenTeams program uses CIE education to provide advanced learning opportunities that can be incorporated into the school day. Faculty sponsors receive training and support from Lemelson-MIT from the time they start the initial application. Teams that move past the first round of applications have instructional time during the school day to learn more about the technical aspects of the design process. The Lemelson-MIT InventTeams program works to reduce excellence gaps by expanding advanced learning opportunities through a university partnership, by giving high school students and their faculty sponsors time during the school day to frontload the content and skills needed for the invention process, and by providing relevant CIE professional learning for faculty sponsors.

Since the program's inception, 12 teams have been awarded patents ("Patents granted" 2021). In 2010, a team from Ardsley High School in New York received a patent for a wheelchair attachment that allows an individual to move over a curb without assistance. In 2016, a team from Natick High School in Massachusetts received a patent for a robot designed to assist first responders in below-ice search and rescue missions. In 2021, a team from Northeast High School in Oakland Park, Florida received a patent for a mosquito agitation device created to inhibit the development of larvae in water sources and prevent the spread of insect-borne diseases (e.g., West Nile Virus). Current teams are investigating a variety of sustainability and safety initiatives. These Lemelson-MIT InvenTeams engage in collaborative processes using technology tools for research, design, and production to create unique mobility, public health, and safety products to meet pressing needs they have identified in their communities.

5.6.2 Purdue EPICS High School Service-Learning

In 1995, Purdue University established the Engineering Projects in Community Service (EPICS) program to engage students in engineering and computing projects that benefit their local communities (https://engineering.purdue.edu/EPICS). The program partners student teams with non-profit organizations in their area to identify a need. Then, they team up with mentors from area universities and corporations to create a project to fill those needs. EPICS is open to elementary, middle, and high school students and it can be implemented within existing courses, as a course itself, or as an extracurricular option for schools. Teams work through the EPICS design process, which begins with project identification and moves through the specification development, conceptual design, detailed design, and delivery phases, with additional stages for maintenance, retirement, and redesign. The service-learning program provides an online community with curriculum resources and instructor support. The Purdue EPICS K-12 program is working to reduce excellence gaps by expanding advanced learning opportunities for students at all grade levels during the school day, by creating community partnerships to secure funding, and by providing training and support for educators facilitating the CIE learning experiences.

EPICS projects address education, access and abilities, human services, and environmental issues within students' home and school communities ("EPICS projects" 2020). One EPICS team from Kyrene Aprende Middle School in Chandler, Arizona created mobile science labs for STEM teachers delivering instruction in classroom spaces not outfitted with traditional lab equipment. A class from Elizabeth Seton High School in Bladensburg, Maryland created an adaptive kicking machine so students with mobility challenges could participate in inclusive physical education activities. The EPICS team from Metro Tech High School in Phoenix, Arizona coordinated with the local public library to design and build concrete chess tables to give youth and families a safe and healthy way to engage with other community members. Students from Eastern Tech High School in Baltimore, Maryland built a prototype of a device that can be placed in a storm drain to collect plastic waste material (e.g., shopping bags, water bottles) to prevent ocean pollution. EPICS classes collaborate with community organizations and professional mentors to use technology tools to conceptualize, design, and share prototypes to solve pressing community issues.

5.7 Conclusion

In a speech at the University of Pennsylvania in 1940, President Franklin D. Roosevelt noted that "we may not be able to prepare the future for our children, but we can prepare our children for the future." At the time of his address, the Second World War had been raging in Europe for over a year. America's course was uncertain, but Roosevelt knew that the way forward was to reevaluate how the next generation was being prepared for the world they would inherit from their parents and grandparents. The COVID-19 global pandemic has created unprecedented challenges for the education system and for students from the many households impacted by food, housing, and job insecurity (Karami and Parra-Martinez 2021). When teachers embrace CIE principles and make them part of the K-12 curriculum and school culture, the next generation can become creative, innovative, and entrepreneurial-minded graduates ready for the challenges of postsecondary education and the workforce. As teachers use technology to bring creativity into the classroom, students can learn valuable skills that will allow them to approach current and future societal problems with effective and lasting solutions. There is no one-size-fits-all method for reforming education, but technological advances have given teachers room to run with creativity, innovation, and entrepreneurship in the classroom to improve how they prepare students for the future. We are optimistic about the possibilities of combining CIE education and integrated technology to work toward equity and excellence.

References

"An introduction to design thinking: Process guide." (2010). Hasso Plattner Institute of Design. Stanford University.

"AP with WE service learning." (2021). College Board. https://apcentral.collegeboard.org/launch-grow-ap-program/collaborations-with-ap/ap-we-service

Beghetto, R. A. (2013). *Killing ideas softly? The promise and perils of creativity in the classroom.* Information Age Publishing.

Business Roundtable. (2018). *Workforce partnership initiative.* https://www.businessroundtable.org/wpi

Callahan, C. M., Plucker, J. A., Gluck, S., & Rodriguez, C. (2020). Inclusion of academically advanced (gifted) students. In J. M. Kauffman (Ed.), *On educational inclusion: Meanings, history, issues, and international perspectives* (pp. 176–194). Routledge.

Cheung, A. C. K., & Slavin, R. E. (2012). How features of educational technology applications affect student reading outcomes: A meta-analysis. *Educational Research Review, 7*(3), 198–215. https://doi.org/10.1016/j.edurev.2012.05.002

Cheung, A. C. K., & Slavin, R. E. (2013). The effectiveness of educational technology applications for enhancing mathematics achievement in K-12 classrooms: A meta-analysis. *Educational Research Review, 9*, 88–113. https://doi.org/10.1016/j.edurev.2013.01.001

Covay, E., & Carbonaro, W. (2010). After the bell: Participation in extracurricular activities, classroom behavior, and academic achievement. *Sociology of Education, 83*(1), 20–45. https://doi.org/10.1177/0038040709356565

Dino, R. N. (2016). Connected but different: Comparing and contrasting creativity, innovation, and entrepreneurship. In J. A. Plucker (Ed.), *Creativity and innovation: Theory, research, and practice* (pp. 23–34). Prufrock Press.

"Europe 2020 flagship initiative: Innovation union." (2011). Publication Office of the European Union. https://op.europa.eu/en/publication-detail/-/publication/440f4722-e9ad-43b2-892a-aba42909c54a

"EPICS projects." (2020). https://engineering.purdue.edu/EPICS/k12/EPICS-K-12/Projects

"From the classroom to the community." (2015, March). IB World. https://ibo.org/ib-world-archive/march-2015-issue-71/from-the-classroom-to-the-community/

Giannakos, M. N., Divitini, M., & Iversen, O. S. (2017). Entertainment, engagement, and education: Foundations and developments in digital and physical spaces to support learning through making. *Entertainment Computing, 21*, 77–81. https://doi.org/10.1016/j.entcom.2017.04.002

Global Business Coalition for Education. (2018). *Preparing tomorrow's workforce for the fourth industrial revolution*. https://gbc-education.org/project/youth-skills/

Grossman, T., Reyna, R., & Shipton, S. (2011). *Realizing the potential: How governors can lead effective implementation of the common core state standards*. National Governors Association. https://files.eric.ed.gov/fulltext/ED532524.pdf

Guo, J. & Woulfin, S. (2016). Twenty-first century creativity: An investigation of how the partnership for 21st century instructional framework reflects the principles of creativity. *Roeper Review, 38*(3), 153–161. https://doi.org/10.1080/02783193.2016.1183741

Hackmann, D. G., Malin, J. R., & Bragg, D. D. (2019). An analysis of college and career readiness emphasis in ESSA state accountability plans. *Education Policy Analysis Archives, 27*(160), 1–28. https://doi.org/10.14507/epaa.27.4441

Henriksen, D., Mishra, P., & Fissure, P. (2016). Infusing creativity and technology in 21st century education: A systemic view for change. *Journal of Educational Technology & Society, 19*(3), 27–37. https://www.jstor.org/stable/jeductechsoci.19.3.27

"ISTE standards for students." (2021). International Society for Technology in Education. https://www.iste.org/standards/iste-standards-for-students

Karami, S., & Parra-Martinez, F. A. (2021). Foolishness of COVID-19: Applying the polyhedron model of wisdom to understand behaviors in a time of crisis. *Roeper Review, 43*(1), 42–52. https://doi.org/10.1080/02783193.2020.1840467

Koretz, D. (2008). The pending reauthorization of NCLB: An opportunity to rethink the basic strategy. In G. L. Sunderman (Ed.), *Holding NCLB accountable: Achieving, accountability, equity, & school reform* (pp. 9–26). Corwin Press.

Lee, L. E., Meyer, M. S., & Crutchfield, K. (2021). Gifted classroom environments and the creative process: A systematic review. *Journal for the Education of the Gifted, 44*(2), 107–148. https://doi.org/10.1177/01623532211001450

Lockhart, K., Meyer, M. S., & Crutchfield, K. (2021). A content analysis of selected state plans for gifted and talented education. *Journal of Advanced Academics*. Advance Online Publication. https://doi.org/10.1177/1932202X211026240

Mishra, P., Fahnoe, C., Henriksen, D. (2013, January). Creativity, self-directed learning, and the architecture of technology rich environments. *TechTrends, 57*(1), 10–13. https://doi.org/10.1007/s11528-012-0623-z

National Association of Colleges and Employers. (2018). *The 2018 student survey report*. https://www.naceweb.org/store/2018/2018-nace-student-survey-report/

No Child Left Behind Act of 2001, 20 U.S.C. § 6319 (2008). https://www.congress.gov/107/plaws/publ110/PLAW-107publ110.htm

Novak, A. M., Lewis, K. D., & Weber, C. L. (2020). Guiding principles in developing equity-driven professional learning for educators of gifted children. *Gifted Child Today, 43*(3), 169–183. https://doi.org/10.1177/1076217520915743

Omdahl, S. N., & Graefe, A. K. (2016). Investing in creativity in students: The long and short (term) of it. In J. A. Plucker (Ed.), *Creativity and innovation: Theory, research, and practice* (pp. 205–221). Prufrock Press.

"P21 framework definitions." (2009). Partnership for 21st Century Skills. https://eric.ed.gov/?id=ED519462

"Patents granted." (2021). https://lemelson.mit.edu/sites/default/files/2021-04/LMIT_Patent_Marketing_Flyer_03.25.21_pdf

Plucker, J. A. (2016). Creative articulation. In J. A. Plucker (Ed.), *Creativity and innovation: Theory, research, and practice* (pp. 151–163). Prufrock Press.

Plucker, J. A., Beghetto, R. A., & Dow, G. T. (2004). Why isn't creativity more important to educational psychologists? Potential, pitfalls, and future directions in creativity research. *Educational Psychologist, 39*(2), 83–96. https://doi.org/10.1207/s15326985ep3902_1

Plucker, J. A., Burroughs, N., & Song, R. (2010). *Mind the (other) gap! The growing excellence gap in K-12 education*. Center for Evaluation and Education Policy.

Plucker, J. A., & Callahan, C. M. (2020). The evidence base for advanced learning programs. *Phi Delta Kappan, 102*(4), 14–21. https://doi.org/10.1177/0031721720978056

Plucker, J. A., Hardesty, J., & Burroughs, N. (2013). *Talent on the sidelines: Excellence gaps and America's persistent talent underclass.*Center for Evaluation and Education Policy.

Plucker, J. A., & Peters, S. J. (2016). *Excellence gaps in education: Expanding opportunities for talented students*. Harvard Education Press.

Plucker, J. A., Peters, S. J., & Schmalensee, S. (2017). Reducing excellence gaps: A research-based model. *Gifted Child Today, 40*, 245–250. https://doi.org/10.1177/1076217517723949

Powell, W. W., & Snellman, K. (2004). The knowledge economy. *Annual Review of Sociology, 30*(1), 199–220. https://doi.org/10.1146/annurev.soc.29.010202.100037

Rinn, A. N. (2020). *Social, emotional, and psychosocial development of gifted and talented individuals*. Prufrock Academic Press.

Sawyer, R. K. (2012). *Explaining creativity: The science of innovation* (2nd ed.). Oxford University Press.

Shalley, C. E., Hitt, M. A., & Zhou, J. (2015). Integrating creativity, innovation, and entrepreneurship to enhance the organization's capability to navigate in the new competitive landscape. In C. E. Shalley, M. A. Hitt, & J. Zhou (Eds.), *The Oxford Handbook of Creativity, Innovation, and Entrepreneurship* (pp. 1–14). Oxford University Press.

Snellman, K., Silva, J. M., Frederick, C. B., & Putnam, R. D. (2015). The engagement gap: Social mobility and extracurricular participation among American youth. *The Annals of the American Academy of Social and Political Sciences, 657*(1), 194–207. https://doi.org/10.1177/0002716214548398

Spoon, R., Rubenstein, L. D., Shively, K., Stith, K., Ascolani, M., & Potts, M. L. (2020). Reconceptualizing professional learning with the gifted field: Exploring the instruct to innovate model. *Journal for the Education of the Gifted, 43*(3), 193–226. https://doi.org/10.1177/0162353230933001

Subotnik, R. F., Olszewski-Kubilius, P., & Worrell, F. C. (2011). Rethinking giftedness and gifted education: A proposed direction forward based on psychological science. *Psychological Science in the Public Interest, 12*(1), 3–54. https://doi.org/10.1177/1529100611418056

"Sustainable Development Goals." (2017). United Nations Department of Economic and Social Affairs (Sustainable Development). https://sdgs.un.org/goals

United Nations Educational, Scientific, and Cultural Organization. (2015). *Education 2030: Incheon declaration and framework for action. Towards inclusive and equitable quality education and lifelong learning for all.* https://iite.unesco.org/publications/education-2030-incheon-declaration-framework-action-towards-inclusive-equitable-quality-education-lifelong-learning/

Venkataraman, S. (1997). The distinctive domain of entrepreneurship research: An editor's perspective. In J. Katz & R. Brockhaus (Eds.), *Advances in Entrepreneurship, Firm Emergence, and Growth* (Vol. 3, pp. 119–138). JAI Press.

Weininger, E B., Lareau, A., & Conley, D. (2015). What money doesn't buy: Class resources and children's participation in organized extracurricular activities. *Social Forces, 94*(2), 479–503. https://doi.org/10.1093/sf/sov071

World Economic Forum. (2020). *The future of jobs report 2020.* https://www.weforum.org/reports/the-future-of-jobs-report-2020

Zheng, B., Warschauer, M., Lin, C., & Chang, C. (2016). Learning in one-to-one laptop environments: A meta-analysis and research synthesis. *Review of Educational Research, 86*(4), 1052–1084. https://doi.org/10.3102/0034654316628645

Jonathan A. Plucker is the Julian C. Stanley Professor of Talent Development at Johns Hopkins University. He is past-president of the Society for the Psychology of Aesthetics, Creativity, and the Arts and the National Association for Gifted Children. He has received the Distinguished Scholar Award from both the Society and NAGC. His work focuses on creativity education, assessment, and policy.

Melanie S. Meyer is a Postdoctoral Research Fellow at Johns Hopkins University in the School of Education and Center for Talented Youth who studies education policy and talent development. She has over 20 years of experience teaching English, Language Arts, and Reading in PK-12 settings. She has designed curriculum and created evidence-based professional learning sessions for educators at the district, state, and national levels. Her research focuses on identity development, school-based talent development opportunities, and how those experiences prepare students for further postsecondary talent development. She holds a B.A. in English from the University of Texas at Austin and a PhD in Educational Psychology from the University of North Texas.

Sareh Karami is an Assistant Professor of Educational Psychology at Mississippi State University. Karami earned her doctorate in Educational Studies from Purdue University. Sareh received her bachelor's and first master's in clinical psychology from the University of Tehran. She served as the head of the Iranian gifted school's research and extracurricular programs department for more than 10 years. Sareh left her job to do more graduate work in education at the University of British Columbia, Canada. She received her second master's in education from UBC. She has developed two theories of wisdom and published several articles on wisdom, creativity, and intelligence.

Mehdi Ghahremani is an Assistant Professor of Educational Psychology at the Department of Counseling, Educational Psychology, and Foundations, Mississippi State University; where he teaches and study human development to promote understanding of differences of people and analyzing effective and productive modes of thinking such as critical thinking, design thinking, and wisdom. He received his Ph.D. in Educational Studies, with a focus on Gifted, Creative, and Talented Studies from Purdue University. He is a mixed-method researcher certified in quantitative and qualitative inquiry. His research interests center around creative thinking and team creativity; STEM-talent development, psychology of design thinking, and wisdom.

Chapter 6
Creativity, Embodiment and *Ensembles* Through Technological Interactions in Critical-Creative Higher Education

Tatiana Chemi

Abstract The global pandemic brought urgent challenges to societies in need of creative approaches to education. As a response to this crisis, technology has been used to provide steady educational offerings to individuals who could not meet in person. Adjustments to this need varied according to the institutions' previous experience with technology-driven education. Those who had established online educational programs well ahead of lockdowns, turned to technology as a tool for surviving through the pandemic. The former case is documented through the development of attitudes towards technology in education, which goes from the understanding of technology as a mere tool for assisting groups in problem-solving, to rethinking technology's entrepreneurial value. In the other response to the pandemic, where technology was not already an educational habit or mindset, problems were solved abruptly, often intuitively and with low technological skills. This research investigates this kind of experience. Looking at creativity literature and its development towards critical-relational perspectives, and the educators' perception of technology, this chapter focuses on the collective side of creativity as mediated through digital education. The key concept investigated is that of critical-creativity in relational exchanges: what happens when educators engage communities of learners in critical-creative practices through technological platforms?

Keywords Critical stance · Distributed collaboration · Technology mediation · Relational creativity · Teaching in a pandemic

T. Chemi (✉)
Department of Culture and Learning and The Faculty of Humanities, Center for Educational Policy Research, Aalborg University, Aalborg, Denmark
e-mail: tch@ikl.aau.dk

D. Henriksen, P. Mishra (eds.), *Creative Provocations: Speculations on the Future of Creativity, Technology & Learning*, Creativity Theory and Action in Education 7,
https://doi.org/10.1007/978-3-031-14549-0_6

6.1 Society in Need of Critical Creativity

The impact of COVID-19 on creative education has brought about the need of a radical change in the ways in which humans think, behave and envision their future in pedagogical ecologies. To imagine and establish new, appropriate models of critical responsibility, it is necessary to act creatively. Creativity becomes, in this context, a fundamental skill, not only instrumental to business innovation, but core to systems where sustainable solutions become necessary to our very survival as humans within ecosystems. According to Adams and Owens (2015) "the conditions for creative practices to flourish in education are largely determined by the extent to which democratic principles are established" (p. 7). The inverse influence might also apply: the conditions for democratic principles to flourish in society are largely determined by the extent to which creative practices are established in education. Pedagogical spaces that are critical and creative operate at the periphery of artistic disciplines and social practices, not to alienate themselves, but to allow for the existence of inquiry and to protect investigative processes that are creative. This is challenging for formal educational institutions that still understand creativity as the product of genius, talent and sudden sparks of inspiration.

The problem addressed in the present chapter is the assumption that higher education is -or has to be- centred on logical-verbal exchanges, individualism, and curricula. This belief is perpetrated to the disadvantage of learners who do not comply with normativity or are socio-culturally excluded from holistic, ethical, engaging developmental opportunities, which is highly problematic for democracy. In critical-pedagogical terms, learners are denied the authority to *perform* their own identity and agency in learning environments shaped on "the banking system of education" (Freire 1996). It is urgent to reframe creativity as fundamentally critical (Adams and Owens 2015), distributed (Glăveanu 2014; Sawyer 2007; Sawyer and DeZutter 2009), relational (Wyatt 2018), ecologic (de Bruin and Harris 2017), multiple and plural (Burnard et al. 2018) and co-creative (Chemi 2018; Chemi and Krogh 2017).

6.2 Technology as Disruption

The world pandemic crisis in 2019–2020 brought new urgent challenges to societies in need of creativity and creative approaches to education. As a global response to this crisis, technology has been used to provide steady educational offerings to individuals who could not meet in person. Adjustments to this need varied accordingly to the institutions' previous involvement with technology-driven education: some had been establishing online educational courses well ahead of lockdowns, others turned to technology as a tool for surviving through the pandemic. The former case is documented through the development of attitudes towards technology in education, which goes from the understanding of technology as a mere tool for assisting groups in cognitive processes of problem-solving (Siau 1995) to means of rethinking technology in its entrepreneurial value (Resnick 2002). Interestingly, for creative education, Zhou et al. (2014) reframe technology as an active collaborator and friendly co-creator. They suggest that IT and creativity have a close and complex relationship: "on the one hand, IT supports and mediates learners with diverse creative learning environments; on the other hand, in order to handle problems required by learning tasks in IT-supported educational contexts, learners need creativity to move on in the learning process. This means the educational role of IT-related disciplines is not only to prepare students with the ability of using IT well, but also to provide a positive attitude towards IT systems and a confident, responsible and creative use of IT in the information society" (p. 455). The paradox of creative education is that, on the one hand, technology can help in designing stimulating learning environments, on the other, IT-driven education requires its users to acquire appropriate knowledge, skills and mindsets. The use of technology in education becomes a creative problem, which demands original solutions.

In the other response to the pandemic, where technology was not already an educational habit or mindset, problems were solved abruptly, often intuitively and with low technological skills. The present research investigates this kind of experience. Looking at creativity literature and its development towards critical-relational perspectives, and the educators' perception of technology, I focus on the collective side of creativity as mediated through digital education. The key concept I investigate is critical-creativity in relational exchanges: what happens when educators engage communities of learners in critical-creative practices through technological platforms?

6.3 Methodology

The main empirical body relies on my autoethnographic reflections (Jones et al. 2016) retrieved within the collaborative digital project *Massive and Microscopic* (Markham et al. 2021). To give space to the students' voices, quantitative and qualitative course evaluations at a higher educational Master programme on creative learning at a Danish university, ethnographic field observations and informal conversations (29–30 October 2020) have been collected and analysed narratively. The Master is a two-year part-time professional development programme for in-service practitioners in several educational areas. Topics and teaching methods draw from creative learning, arts-based education (Barone and Eisner 2011), artist-led learning (Chemi et al. 2015) and theories and methodologies related to the understanding of educational creativity as deeply embodied, relational and critical. Ethical measures have been taken in concealing the students' names with anonymity and by careful considerations in the sampling of autoethnographic narratives that are respectful to others.

In this contribution, I understand creativity as socially and culturally situated and as the outcome of a shared commitment to the creation of what is new and appropriate in a given context: creativity can be seen as the "interaction among *aptitude, process, and environment* by which an individual or group produces a *perceptible product* that is both novel and useful as defined within a *social context*" (Plucker et al. 2004, p. 90). I propose the concept of *ensemble creativity*, which is investigated by means of a desk-study and preliminary review looking at terms related to 'ensemble': group creativity, team creativity, collaboration, participatory creativity, distributed creativity, creative environments, milieu, *press*. Beyond discourses of talent, genius and of high achieving individuals, I advocate for a creative mindset in everyday life that is meaningful within a community, possibly life-changing, positively pedagogic and critical, and where technology has the role of disruptive co-creator.

6.4 Pandemic Creativity

Educators who bring creativity into the classroom tend to work hard (Harris 2016). Several obstructions stand in their way, but creativity is also about overcoming challenges, so creative problems can be experienced as motivating (Chemi et al. 2015). What is more creatively inspiring than a world crisis that takes away all the taken-for-granted possibilities for creative education? Educators who make use of bodily, arts-based, sensory, dramaturgical, narrative tools to shape a critical-creative classroom got lost during the COVID-19 suspension of physical classes and all their sensory practices. I described elsewhere (Chemi 2021) how the pandemic "introduced more than a creative obstruction to my teaching plan: it was a disaster. It felt like a thorough obliteration of my groundwork, of the developed materials, and of the very foundation of my teaching philosophy" (p. 854). The shock was a wake-up call and a necessary opportunity to rethink my teaching method within the frames of digital communication. Technological solutions felt, back in March 2020, like a defeat for the creative method of my choice and for the very content of my classes, that is, history of creativity, arts-based methods in education and educational research, the role of emotions in education, creative writing and theoretical/methodological perspectives related to these topics. Because of my lack of previous experiences with technology-driven teaching, the abruptness of the change required and the intensity of the teaching schedule (three teaching weeks, all-day-long classes, evening teaching), I was torn between despair ("it is impossible", Chemi 2021, p. 853) and stubbornness ("this situation would *not* be allowed to mess with my creative intentions!", Chemi 2021, p. 854). This is well described in Henriksen et al. (2016), where teachers' beliefs are mentioned as fundamental to the extent in which technology can become a creative element in education. They also address the opportunities I encountered: on the one hand, "to creatively repurpose existing tools for [my] educational purposes" (Henriksen et al. 2016, p. 31), on the other, to rethink my representation of technology and offer the students new opportunities for exploring creative processes. My way out of this crisis was a collaborative, technology-mediated, autoethnographic project of self-reflection on my creative teaching (Markham et al. 2021). This brought me to the conceptualisation of *ensemble creativity*.

6.5 Individuals in Groups

Too often have collective and critical perspectives on creativity and creative education been excluded in literature. Since Guilford's address to the American Psychology Association in 1950 (Guilford 1950), creative skills for education and work-life have been valued as fundamental to growth. Creativity skills are still among the top three most desired skills for the future (World Economic Forum 2018), because they are expected to boost economic and techno-industrial development. A different approach aims at disclosing uncritical discourses about the social relevance of creativity, bending them towards human growth and community development. Craft (2001, 2005) investigated the educational potential of creativity through conflicting discourses and Harris (2016) has made explicit that the "creative turn" (p. 17) in education is and can be a critical act. The understanding of creativity as relational and collective has been a conquest only relatively recently achieved in creativity studies. With its origins in romantic and industrialist mindsets, the idea that creativity is the achievement of individual talent and genius has been influencing centuries of studies (reviewed in Kaufman and Sternberg 2010). However, this perspective radically changed when creativity began to be seen as collective-relational and distributed in time and space (Glăveanu 2014; Sawyer 2007; Sawyer and DeZutter 2009).

Even within more aware studies, collaborative aspects of creative processes hide behind individualist and anthropocentric ideals. Stein (1974, 1975) studied the impact of environment on individual creativity by reviewing studies up until the mid 1970s. His understanding of group creativity strongly relies on the individual as the emanating centre of creativity, even though he looks at group level, pressure and the use of a range of embodied techniques, including role play. Ryhammar and Brolin (1999) explain the environment's influences through psychological criteria (personality, cognition, training) that are mainly behavioural. Csikszentmihalyi (1997), Gardner (1993) and Amabile (1983) investigate organisations and systems by focusing on individuals, whose motivation is seen as essential to any creative initiative.

6.6 Creativity in Education: The Collective Level

In Craft (2001), where criteria of educational achievement are reframed to include creativity, collaborative skills and teamwork are both mentioned as necessary pre-requisites for and effect of creative work. Shaheen (2010) points to two opposing attitudes towards the application of creativity in education. On one hand is the conformist and conforming school that manages to overindulge students or even ‘kill’ their creativity (Amabile 1998; Robinson 2011). On the other hand, humanistic psychology is taken as an example of educational aim towards free and original thinkers. Several reconceptualisations of creativity, education and their relationship mostly advocate for a better, more inclusive presence of creativity in formal education (Craft 2001; Burnard et al. 2018).

Grodoski (2015) notices a renewed interest from educational institutions towards creativity, however collaborations and cross-disciplinarity are not only poorly taken into consideration, but are also not successfully practised. The field that has demonstrated more curiosity about the pluralistic elements of creative processes is business studies, probably due to a renewed attention to the role of the arts in organisational change (Chemi and Du 2018) and leadership development (Taylor 2012; Taylor and Ladkin 2009). Education is seen to have a role in the promotion of creative skills and in the advancement of collaborative practices. This is claimed to prepare the creative and competitive learners needed to work in knowledge organisations (Craft 2001).

Against the crisis of creativity in education, Clapp (2017) counterpoises the understanding of creativity as participatory, collaborative and distributed: “participatory creativity combines individual agency with sociocultural models of creativity, emphasizing the roles that people play in change. [A] participatory reframing of creativity opens up engagement in invention and innovation to everyday practice, as opposed to the extraordinary feats of ideation favoured by more eminence-based models of creativity” (p. 32). Clapp brings as examples design and makerspaces, but the same characteristics are common to artists’ studios and laboratories. Differently, theatre laboratory opens to pedagogical spaces that imply the making of autonomous dramaturgies by means of meaningful and meaning-generating processes (Chemi 2018).

6.7 Paradigm Shift

Vera John-Steiner (2000) studies creative collaborations in small and medium-size teams, showing that creative ideas travel amongst individuals and milieus through distributed trajectories. Regardless of context and type of collaboration, partnerships of creative individuals build on the shared engagement in active conversations amongst their members who are close in time and space (collaborators) but also with individuals who are distant in space and in time (role models, inspirers, teachers). She relies explicitly on Vygotskian cultural-historical conviction that "creative activities are social, that thinking is not confined to the individual brain/mind, and that construction of knowledge is embedded in the cultural and historical milieu in which it arises" (John-Steiner 2000, p. 5). Chadwick and de Courtivron (1993) explore small-size collaborations considering the relational and dialogic, looking at the constructivist concept of *significant others* in intimate relationships, drawing from creative collaborations between artists romantically engaged with each other. Melnick and Nevis (2003) look at creativity in long-term intimate relationships between artists emphasising the role that *form* is given in the artists' creative processes, as a sort of third dialogic partner and locating the creative couples' flourishing strategies in the following: "creative adjustment", "experimental attitude", "hard work and discipline", "creative stabilization", "bearing disappointment", "maintaining interest", "sense of humor" (p. 228). John-Steiner (2000) insists on the *interdependence* of individuals within creative collaborations in and on the *distributed* character of creative exchanges:

> Distributed collaboration is widespread. It takes place in casual settings and also in more organized contexts. These include conversations at conferences, in electronic discourse communities, and among artists who share a studio space. In these groups, participants exchange information and explore thoughts and opinions. Their roles are informal and voluntary. [...] The participants in distributed collaborative groups are linked by similar interests (John-Steiner 2000, pp. 197–198).

Where in John-Steiner (2000) distribution is one of the different shapes that creative collaborations may take in a continuous adjustment of roles, values, working methods and patterns (p. 197), in Sawyer (2007) we find the definition of "group genius". Sawyer (2007) relies on the observation of creative collaborations that occur "over many years of interactions, trial and error, and false starts" (Sawyer 2007, p. 9), obliterating the myth of creativity emerging from sudden sparks of inspiration that are directly put in *production*. Sawyer's (2007) portrait of the creative individual evolves in a collaborative network of influences that need time to emerge and are based on the group's members' commitment to building on each other's ideas and capacity for deep listening, suspension of judgment (meaning becomes clear afterwards), and self-organising processes (bottom-up). Individual members of the creative group actively become open to surprising results, while still aware that "innovation is inefficient" (Sawyer 2007, p. 16). Glăveanu (2014) and Clapp (2017) argue that group genius might hardly replace the individual genius perspective with collective genius. In fact, Sawyer (2007), in his early portrait of creative collaborations, strongly relies on psychological traits, ending up by applying flow theory on group dynamics. Later, Sawyer and DeZutter (2009) argue that merely looking at individual mental processes is inaccurate, especially in the case of collaborative groups. They propose the concept of distribution, borrowed from cognitive science, and apply it to relational processes. Lea and Nicoll (2002) introduced the concept of distributed learning to the field of technology-enhanced learning environments when technological progress made it possible to connect distant learners globally, creating the need to rethink education as distributed in space and time. Miettinen (2006) and Sawyer and DeZutter (2009) define *distributed creativity* as emerging in "situations where collaborating groups of individuals collectively generate a shared creative product" (p. 82) against the background of shared cultural practices and social interactions. According to Glăveanu (2014), distributed creative exchanges stretch beyond any limitations in time and space, allowing for conversations across *chronos* and *topos*: "the production of what is creative is distributed both between people and across time" (p. 25).

Recent trajectories move away from psychological accounts and engage in the collective action of "transversal ensembles" (Braidotti 2019, p.182). Creativity is understood as socially and culturally situated as the "quest for affirmative alternatives" (Braidotti 2013, p.165) that are critical, visionary and ethical. This situatedness is non-linearly entangled (Barad 2007) with the materiality of the human and non-human (including technology) in reciprocal influences. Theory is reframed as critical cartography of embodied and participatory practices "aimed at the creation of new concepts and affirmative values" (Braidotti 2019, p.146). Creative criticality makes use of "radical epistemologies" (Braidotti 2019, p.147) that reject othering, naturalisation, racialisation, and commit to ethical "transversal assemblages" (Braidotti 2019, p.140), thus becoming active means for systemic change.

6.8 Ensemble Creativity

Including technology in creative processes as a dialogic partner or co-creator (Latour 1987) shifts the focus of education towards new ways of looking at learning environments and relationships. I propose a shift to conceiving critical-creative collaborations, by looking at *ensemble creativity*. Collaborative creativity relies on affects: group flow (Sawyer 2007) or team flow (Van Hout et al. 2016), co-creative ensembles (Chemi and Pompa 2017), relational creativity (Wyatt 2018). An example of this *affective turn* might be Turner and Campbell's (2018) investigation of the creative processes occurring within theatre ensembles, theatre laboratory and networked theatre groups and looking at how these communities engage in radical "generosity" (p. 58). This sustains the socio-political imagining and establishing alternative theoretical/ethical representations of new relationships in social ecologies and diverse/excluded communities, which calls for thinking critically and acting creatively. Creativity becomes a competence, core to democratic participation that can empower learners in reclaiming their own agency through collective action, a form of *ensemble* performance. Knowledge is needed on the ways in which performativity emerges in ensembles, defined as "a group of supporting [performers]" "producing a single effect" (Merriam-Webster 2019) together (French *ensemble* = together). Here, *ensemble* is intended as a group whose participants share interest in and commitment to collaborative art-experiences that are performed, embodied and transformative in contexts that include the technological ones. According to Spry (2016), performance is always relational: the "coperformativity" of the actor's body makes the *ensemble* an original field of empirical inquiry and theory development. However, studies investigating the relationships between education and creativity have mostly ignored ensembles as educational spaces, even though theatre as community intervention has been practised in political and critical theatre and is conceptually sustained by critical pedagogy (Freire 1996; Boal 2000).

John-Steiner's (2000) review of the Group Theatre's history, with its foundation in New York in the thirties and its character of "artistic organism" (p. 88), describes the ensemble's organisation by the shared values of democratic co-participation. Living by their own ideals, its members dealt with contextual challenges finding appropriate organisational and artistic solutions to emerging needs. The Group Theatre's collaborative exchanges might be considered a theatre laboratory (Chemi 2018) even though the ensemble defines itself as *group*. Burgoyne (2018) collects contributions on the topic of drama education, addressing the intersection of creativity, theatre and education, by bringing several perspectives. Monk et al. (2011) explicitly use ensemble as the educational design unit for establishing open space learning. Despite not defining ensemble, the programmes seem to enhance inclusion, effectiveness and engagement in the participants. However, more studies on the ways professional ensembles create and learn would be necessary. This is even more relevant in digital learning environments, where learners can be physically separated from each other and prevented from shaping critical-creative ensembles.

6.9 Massive and Microscopic

In April 2020, to my surprise, I happened to become part of a digital ensemble. When my Spring-term-rush was over, I responded to the call for participation in a distinctive writing project: the *Massive and Microscopic* (Markham et al. 2021), a 21-day collaborative reflection on the ways in which the pandemic was affecting relationships and practices. During the project, participants would receive a daily reflexive prompt to which to respond creatively in multimedia. In my response to prompt #8, the need for distributed companionship clearly emerges:

> Did I miss words? Not necessarily. I was missing a friend, a companion. Something to break the claustrophobic solipsism of the self-reflective room. I needed to invite the world in (M & m 27 May 2020).

This offered an opportunity to reflect on the ways in which I had addressed my teacher's creativity by translating arts-based educational methods into digital frames, and also a platform where distributed and collaborative creativity was practiced by means of technological communication. It gave me the navigation tools to balance the autoethnographic self, with its tantalising voice, and the experiences of the Other. Ideas, intuitions, perspectives started to intermingle by means of low-technological tools: social media groups, Google drive, emails for sharing and communicating, together with video, audio and photo software for the creation of artworks or multimodal responses. Several participants worked with similar associations, built on each other's experiences, and responded emotionally to the shared pieces of reflection. It was like being part of a collective mind, heart and body, a sort of *ensemble* working creatively and critically through technology.

This project made me mindful of the trial-and-error learning harvested in the digital classroom. Early in the digitalisation of my courses, one colleague made me aware of the positioning of the camera for online synchronous teaching or meetings. Rule number one is the positioning of the camera at eye level. I had never thought about that before. As trivial as it may seem, attention to the camera angle offers a specific eye-perspective to the students on the other side and ends up having a relevant role on the digital relationality. When I grabbed the first objects that could function as a base, this was a pile of books. What a beautiful metaphor. Standing on women's shoulders. However, this platform was not very stable and was eventually shifted to a portable record player. What did this shift mean?

> Probably nothing else than the daily interaction of human-non-human even when technological objects are not used in their proper function. They are there, we grab them. They live with us, we live with them. The interaction is nothing to write home about, it is just there. Available and open to disruption. By using this object as a base to my laptop, I disrupt its original function. I exercise my creative right to invent a new appropriate application of the object to my emerging needs (M & m 28 May 2020).

Looking back at my own creativity in teaching digitally, this did make explicit concrete undertakings and revealed my affective reaction to the task. By means of a situational map methodology (Clarke et al. 2015) I was able to retell my experience in actantial terms (Wang and Roberts 2005) by means of clearly emerging antagonistic versus helping factors. The quality of the agentic reaction is described as an immediate ("just do it", this and the following quotes are from M & m 29 May 2020) restructuring of educational materials free of reflection. Even though the organisational demands of my institution stand out clearly in my analysis, what has my attention is my own need to act uncompromisingly, according to my core pedagogical and ideological values, and according to my teacher's personality and creative practice (to "challenge myself"). The challenges met in the process point to the lack of appropriate technological tools, the bureaucratic IT-support that shaped confusion in its users and frustrations due to the lengthy processes, the lack of time for experimentation and informed decision-taking, and my own resistance to technological solutions. The affective response is so strong that some of these challenges provoked a "void" that made me "almost give up" the creative task. On the side of helping factors the following are tracked in my map:

- Colleagues
    - Human facilitator who takes time to guide and help others
    - Modelling of solutions: it was an "eye-opener" and a new experience in an environment where scholars "hold their cards 'tight'". The feelings connected to this were gratefulness, humility and motivation to work hard, in order to honour the colleagues' time spent on helping
    - Understanding and compassion: all being in the same situation, colleagues offered to help and spent time to do so ("it felt 'legitimate' to offer help without offending anyone")

- Knowledge sharing
  - Focused on practical problem-solving and applicable 'how to'
  - Clear, realistic, well-thought frames for sharing
- Organisational attitude
  - Results and actions were not pushed and the educators' readiness in responding to the digital challenge were looked upon with gratefulness
  - Tenure makes it a little easier to navigate in educational changes that can become unsuccessful
- Students
  - Compassion towards learners who were deprived of their rights to education
  - Necessity of understanding the students' needs.To my surprise, the voices and bodies of my students were absent from my logbooks. They were the motivation for all the effort of rethinking teaching and reflecting on it. They were distant partners of my reflections, distributed collaborators in our virtual ensemble. Why were their voices missing? This acknowledgement brought me to analyse the students' course evaluations, in light of my emerging conceptualisations of critical-creativity in ensembles that can be technologically mediated.

6.10 Students' Voices

Responding to course evaluations is seldom a task prioritised by my students on the Master programme in creative learning processes. This year, the standard evaluation form was sent out on 22 June 2020 to 16 students, and a total of 11 chose to participate in the survey, which gives a response rate of 69% versus last year's response of 31%. There seems to be widespread understanding of the urgency to listen to students' responses considering the precarious COVID-19 situation. Not only did the students take seriously the teachers' need for feedback by participating in the evaluation, but they also were very specific about what made the teachers' digital solutions successful or not. Moreover, they specifically addressed the challenge of transforming creative embodied teaching into technological-driven platforms. Subsequent informal conversations disclosed the meaningfulness of this reflection for the students, who are themselves teachers or facilitators of creative learning for a number of different target groups. The overall satisfaction with the course is tainted by the pandemic's effect on teaching, which is described in great detail.

The students express their appreciation for the teachers' ability to react quickly to the closure of physical classes and to complete creative teaching despite the unwanted obstruction. Nevertheless, the students notice how much the teachers' readiness and preparation affect the teaching and their experience of it: on one hand, they are grateful for the great result, on the other, they remark on a different experience when the teacher did not have enough preparation time. Some students are very critical of the digital solution, which they feel contradicts the Master's pedagogy, to the extent that some even considered dropping out:

> The gain of the creative elements was reduced [in the digital form], which was certainly not the fault of the teachers, but of the situation. I will consider whether I will complete next semester if it becomes exclusively digital teaching again, since that part has been disappointing. One of my goals with this education is that I must develop the bodily creative element in theories teaching (anonymous student evaluation, 2020).

Some students point to the frustrations that technology can cause if it does not work: the effectiveness of digital teaching does encompass optimal reflection, critical thinking and creativity, even though some platforms are mentioned as better facilitators than others. What is mentioned as positive is what gives the feeling of experiencing almost the 'real thing', by having the possibility of looking at all the participants at once, the easy trespassing to group work and plenum, and the involvement in activities. Even when the sensory experience is triggered by analogous feelings (sight, conversations, liminal trespassing) in digital teaching there is no time for not understanding and tinkering. Therefore, it is felt to be more tiring and the teacher's work is perceived as the product of a bigger effort and demand. On the positive opportunities offered by technological solutions, the students mention their surprise in finding out that artistic, collaborative and sensory tasks could be done online, such as a collaborative collage, dramatisations, virtual museum tours, artwork analysis, but also the fact that they could hear the recorded lectures again and again. The social environment has been assessed as intensely amputated, which is aggravated by the programme's educational philosophy, the problem- and project-based learning (PBL) (Kolmos et al. 2004) that strongly relies on the students' close collaboration and group-work.

The negative feedback seems to be much less necessary for technology-mediated supervision. Students at this Master are geographically spread and the fact of being able to reach out to one's supervisor in flexible and dialogical ways was positively experienced. Video conversations made supervision smooth and the students did not miss physical presence. On the contrary, classroom teaching with creative topics and teaching methods is compared to 'live' teaching to the disadvantage of the digital solution, not only for the fact of not being able to be physically together in conversations, but also because the students missed the ability "to feel exercises and innovative and creative teaching designs on [their] own bodies, and to collect experiences and opportunities for development in relation to theory but not only primarily [by means of] theory" (anonymous student evaluation, 2020).

The student evaluations point specifically to abundant development opportunities on the digital, by sharing the same curiosity as the teachers: how do we teach creatively on screen? The solution they mention as most interesting is the blended learning or flipped classroom, which could combine the best of both models. It is worth reflecting upon how these technology-integrated solutions can support embodied and critical creativity, together with the relational part of the learning experience. In educational models such as the PBL or participatory creative learning (Clapp 2017), it is fundamental to find ways in which the *ensemble* of students can build up a community of creatives. Are on-screen social activities appropriate to this task? How could the distributed model of creativity be applied to education to model future embodied and critical-creative exchanges?

6.11 Future Trajectories

Summing up on my journey through technology-driven creative teaching in times of pandemic emergency and the creativity theories that support embodied and critical creativity, I can point to a few research directions for the future.

Firstly, the potential for framing creativity as a pedagogical strategy to teach criticality and shape democratic change, also in technology-driven education, still needs to be fully investigated. The future of creativity studies ought to explore in depth these directions both empirically and theoretically, for instance by means of imaginative conceptual frames like *ensemble creativity*. Of use to this end could be new materialist (Barad 2007) and post-human (Braidotti 2013) perspectives, to construct a new framework that removes the (singular) human from the core of creative processes and builds up to the understanding of creativity at the intersection of digital technologies, informal learning and distributed creativity. Multidisciplinarity can be sustained in relational collectives that are felt and perceived as *ensembles*: a togetherness of diverse participants in a shared, multi-modal and affirmative group that is characterised by embodied, sensory and artistic expressions. Creative collectives have the potential to explore the poetic and disruptive potential of technological communication and pedagogy. Relational creative processes experienced as an ensemble can perhaps become "environment[s]-in-life" (Barba 2010, p. 206) in entangled (Barad 2007) exchanges with the world. The students' voices, documented in ordinary evaluation forms, should have been captured differently in the extraordinary situation of pandemic teaching and should have been the building stone of educational institutions' future education policies. How can higher education environments foster ensembles of critical-creative learners and educators, in analogical/digital ways, or in technology-mediated ways that retrieve the embodied creativity and criticality of relational exchanges? The story of how this will happen by including embodied and critical creativity is still to be written and the post-pandemic creative policies will be the platforms on which this plot will unfold. It is to be hoped that the lessons learned will be creatively integrated in future higher education and its approach to technology.

References

Adams, J., & Owens, A. (2015). *Creativity and Democracy in Education: Practices and politics of learning through the arts*. Routledge.

Amabile, T. M. (1998). *How to kill creativity* (Vol. 87). Boston, MA: Harvard Business School Publishing.

Amabile, T. (1983). The Social Psychology of Creativity. New York, NY: Springer-Verlag.

Barad, K. (2007). *Meeting the Universe Halfway: Quantum Physics and the Entanglement of Matter and Meaning*. Duke University Press.

Barba, E. (2010). *On Directing and Dramaturgy. Burning the House*. London/New York: Routledge.

Barone, T., & Eisner, E. W. (2011). *Arts based research*. Sage.

Boal, A. (2000). *Theater of the Oppressed*. Pluto Press.
Braidotti, R. (2013). *The posthuman*. John Wiley & Sons.
Braidotti, R. (2019). *Posthuman knowledge*. John Wiley & Sons.
Burgoyne, S. (Ed.). (2018). *Creativity in Theatre: Theory and Action in Theatre/Drama Education* (Vol. 2). Springer.
Burnard, P., Dragovic, T., Jasilek, S., Biddulph, J., Rolls, L., Durning, A., & Fenyvesi, K. (2018). The Art of Co-Creating Arts-Based Possibility Spaces for Fostering STE (A) M Practices in Primary Education. *In* Chemi, T., & Du, X. (Eds.). *Arts-based methods in education around the world* (pp. 247–281). River Publishers.
Chadwick, W., & de Courtivron, I. (1993). Significant Others: Creativity and Intimate Partnership.
Chemi, T. (2018). *A theatre laboratory approach to pedagogy and creativity: Odin Teatret and group learning*. Springer.
Chemi, T. (2021). It Is Impossible: The Teacher's Creative Response to the Covid-19 Emergency and Digitalized Teaching Strategies. *Qualitative Inquiry*, 27(7), 853–860.
Chemi, T. & Du, X. (Eds.). (2018). *Arts-based Methods and Organisational Learning: Higher Education around the World*. Series: Palgrave Studies in Business, Arts and Humanities.
Chemi, T., & Krogh, L. (Eds.). (2017). *Co-creation in Higher Education: Students and Educators Preparing Creatively and Collaboratively to the Challenge of the Future*. Springer.
Chemi, T., & Pompa, P. (2017). Theatre as co-creative space and as inspiration for higher education. In *Co-Creation in Higher Education* (pp. 99–116). SensePublishers, Rotterdam.
Chemi, T., Jensen, J. B. & Hersted, L. (2015). *Behind the Scenes of Artistic Creativity: Processes of Learning, Creating and Organising*. Frankfurt am Main: Peter Lang.
Clapp, E. P. (2017). *Participatory creativity: Introducing access and equity to the creative classroom*. Routledge.
Clarke, A. E., Friese, C., & Washburn, R. (Eds.). (2015). *Situational analysis in practice: Mapping research with grounded theory* (Vol. 1). Left Coast Press.
Craft, A. (2001). An analysis of research and literature on creativity in education. Report prepared for the Qualifications and Curriculum Authority. Retrieved 20 May 2019 at www.creativetallis.com/uploads/2/2/8/7/2287089/creativity_in_education_report.pdf
Craft, A. (2005). *Creativity in schools: Tensions and dilemmas*. Routledge.
Csikszentmihalyi, M. (1997). Flow and the psychology of discovery and invention. *HarperPerennial, New York, 39*.
de Bruin, L. R., & Harris, A. (2017). Fostering Creative Ecologies in Australasian Secondary Schools. *Australian Journal of Teacher Education*, 42(9).
Freire, P. (1996). Pedagogy of the oppressed (revised). *New York: Continuum*.
Gardner, H. (1993). *Creating Minds: an Anatomy of Creativity Seen Through the Lives of Freud, Einstein, Picasso, Stravinsky, Eliot, Graham and Gandhi*. New York: Basic Books.
Glăveanu, V. P. (2014). Distributed Creativity: Thinking Outside the Box of the Creative Individual. Springer, Cham/Heidelberg/New York/Dordrecht/London. ISBN 978-3-319-05433-9.
Grodoski, C. (2015). Creativity, Policy, and Practice in Three States: An Exploration of Definitions of Creativity among State Art Education Policies, the Life Contexts, and Professional Practice of Middle Level Art Educators. *Marilyn Zurmuehlen Working Papers in Art Education*, 2015(1). Retrieved 18 July 2019 at https://ir.uiowa.edu/cgi/viewcontent.cgi?referer=https://www.google.com/&httpsredir=1&article=1488&context=mzwp.
Guilford, J. P. (1950). Creativity. *American Psychologist, 5*(9), 444–454.
Harris, A. (2016). *Creativity, Education and the Arts*. London, UK: Palgrave Macmillan.
Henriksen, D., Mishra, P., & Fisser, P. (2016). Infusing creativity and technology in 21st century education: A systemic view for change. *Educational Technology & Society*, *19*(3), 27–37.
John-Steiner, V. (2000). *Creative Collaborations*. Oxford: Oxford University Press.
Jones, S. H., Adams, T. E., & Ellis, C. (Eds.). (2016). *Handbook of autoethnography*. Routledge.
Kaufman, J. C., & Sternberg, R. J. (Eds.). (2010). *The Cambridge handbook of creativity*. Cambridge University Press.

Kolmos, A., Fink, F. K. & Krogh L. (Eds.). (2004). *The Aalborg PBL model: progress, diversity and challenges*. Aalborg: Aalborg University Press.
Latour, B. (1987). *Science in action: How to follow scientists and engineers through society*. Harvard University Press.
Lea, M. R., & Nicoll, K. (Eds.). (2002). *Distributed learning: Social and cultural approaches to practice*. Routledge.
Melnick J. & Nevis S.M. (2003). Creativity in Long-Term Intimate Relationships. In M.S. Lobb & N. Amendt-Lyon (Eds.) *Creative License* (pp. 227–238). Vienna: Springer.
Markham, A. N., Harris, A., & Luka, M. E. (2021). Massive and Microscopic Sensemaking During COVID-19 Times. *Qualitative Inquiry*, 27(7), 759–766.
Merriam-Webster. (2019). *Ensemble*. Retrieved 20 August 2019 at www.merriam-webster.com/dictionary/ensemble#h1
Miettinen, R. (2006). The sources of novelty: A cultural and systemic view of distributed creativity. *Creativity and Innovation Management*, 2, 173–181.
Monk, N., Rutter, C. C., Neelands, J., & Heron, J. (2011). *Open-space learning: A study in transdisciplinary pedagogy*. A & C Black.
Plucker, J. A., Beghetto, R. A. & Dow, G. T. (2004). Why Isn't Creativity More Important to Educational Psychologists? Potentials, Pitfalls, and Future Directions in Creativity Research. *Educational Psychologist*, 39(2), 83–96.
Resnick, M. (2002). Rethinking learning in the digital age. In G. Kirkman (Ed.), *The global information technology report: Readiness for the networked world*. UK: Oxford University Press.
Robinson, K. (2011). *Out of our minds*. Tantor Media, Incorporated.
Ryhammar, L. & Brolin, C. (1999). Creativity Research: historical considerations and main lines of development. *Scandinavian Journal of Educational Research*. 43(3), 259–273.
Sawyer, R. K. (2007). *Group Genius: The Creative Power of Collaboration*. New York: Basic Books.
Sawyer, K. R., & DeZutter, S. (2009). Distributed Creativity: How Collective Creations Emerge From Collaboration. *Psychology of Aesthetics, Creativity, and the Arts*, 3(2), 81–92. American Psychological Association.
Shaheen, R. (2010). Creativity and Education. *Creative Education*, 1(3), 166–169.
Siau, K. L. (1995). Group creativity and technology. *The Journal of Creative Behavior*, *29*(3), 201–216.
Spry, T. (2016). *Autoethnography and the Other: Unsettling Power Through Utopian Performatives*. New York & London: Routledge.
Stein, M.I., Stimulating creativity, Vol. 1, Individual procedures, Academic Press, New York, NY, 1974.
Stein, M.I., Stimulating creativity, Vol. 2, Group procedures, Academic Press, New York, NY, 1975.
Taylor, S. (2012). *Leadership craft, leadership art*. Springer.
Taylor, S. & Ladkin, D. (2009). Understanding Arts-based Methods in Managerial Development. *Academy of Management Learning and Education*, 8(1), 55–69.
Turner, J. & Campbell, P. (2018). Radical Care: Performative Generosity and Generativity in Third Theatre. *Performance Research*, 23(6), pp. 58–64.
Van Hout, J. J., van der Borgh, M. & Weggeman, M. C. D. P. (2016). Application of Team Flow Theory. In L. Harmat, F. Ørsted Andersen & J. Wright (Eds.). *Flow Experience: Empirical Research and Applications*. Springer.

Wang, Y., & Roberts, C. W. (2005). Actantial analysis: Greimas's structural approach to the analysis of self-narratives. *Narrative Inquiry, 15*(1), 51–74.

World Economic Forum (2018). *The Future of Jobs Report 2018*. Cologny/Geneva: WEF. Retrieved 7 June 2019 at http://www3.weforum.org/docs/WEF_Future_of_Jobs_2018.pdf.

Wyatt, J. (2018). *Therapy, Stand-Up, and the Gesture of Writing: Towards Creative-Relational Inquiry*. Routledge.

Zhou, C., Chen, H., & Luo, L. (2014). Students' perceptions of creativity in learning Information Technology (IT) in project groups. *Computers in Human Behavior, 41*, 454–463.

Tatiana Chemi, PhD is Associate Professor at Aalborg University, in the Department of Culture and Learning and The Faculty of Humanities, at the Center for Educational Policy Research at Aalborg University. Her major interest is the artistic learning and creative process. She has degrees in Italian Literature and Theatre and a PhD in Contemporary Theatre History. Her work focuses on cross-cultural artistic creativity, arts-integration in higher education and theatre laboratory. She has authored multiple books and articles as well as being the founder and co-editor of the book series The Arts, Creativities and Learning Environments in Global Perspectives at Brill/Sense.

Chapter 7
The Future of the Fine Arts

Aaron Kozbelt

Abstract Across human cultures, artistry and aesthetic behaviors are pervasive and varied, but they also show some underlying similarities. Socio-cultural explanations emphasize cross-cultural variation; psychobiological explanations emphasize commonalities, often attributing aesthetic regularities to evolutionary forces. Building on a psychobiological foundation, I argue that just as finished artworks inform human aesthetics, systematic patterns of trans-historical changes within fine arts traditions do so as well. Amply documented trans-historical trends show that creators continually seek critical attention through originality. A rampant drive for novelty eventually runs counter to evolutionarily ingrained aesthetic biases, making artworks less effective at communicating. This tendency is evident in elite art forms in recent generations, leading to a postmodern 'crisis' within those traditions. Speculating on future developments in the arts, I argue that human nature will continue to strongly constrain the arts, especially in their more popular forms; however, when technology arises that fundamentally usurps or supplants human nature, all bets are off about what the arts will be like, what functions they may serve, and whether something akin to the arts even need endure.

Keywords Creativity · Evolution · Aesthetics · Art history · Culture · Postmodernism

The fine arts number among humanity's greatest achievements. For millennia, creative productions in aesthetic domains – like painting, sculpture, music, imaginative writing, dance, and architecture – have been a source of wonder, instruction, and solace for the members of our species. Beyond their cultural importance and experiential impact, especially in recent generations the arts have also showcased striking – even bewildering – variation in media, styles, conceptions, and social

---

A. Kozbelt (✉)
Psychology Department, Brooklyn College, Brooklyn, NY, USA
e-mail: AaronK@brooklyn.cuny.edu

D. Henriksen, P. Mishra (eds.), *Creative Provocations: Speculations on the Future of Creativity, Technology & Learning*, Creativity Theory and Action in Education 7,
https://doi.org/10.1007/978-3-031-14549-0_7

functions. Throughout this chapter, I argue that understanding the sources and limits of this aesthetic variability provides a window into human nature itself.

This understanding, rooted in the history and current condition of the arts, also has implications for forecasting our future and the role of the arts in that future. Arguably, the artistic environment today is different from any other in our species' history. The arts themselves are pluralistic in available media and styles, play out within a labyrinth of aesthetic niches, and are theoretically self-conscious to an unprecedented degree. They function in social settings that are increasingly globalized and distributed, technologically embedded, and informationally interconnected as never before.

This situation raises questions about the relations among human nature, the arts, society, and technology – past, present, and future. For instance, what do the legacies of our evolutionary and cultural histories, coupled with our novel contemporary environment, tell us about the connection between the arts and human nature? How much is human artistry constrained by our biological inheritance, versus being a blank slate of open-ended possibility? To the extent that the arts reflect human nature, when the arts change, does human nature change? As we anticipate the extension of the arts into the future, what will be more important, the apparent diversity of human artistry or what the arts appear to have in common? What is the ongoing role of technology in the arts? Can one make any meaningful predictions about the prospects for the fine arts into the future – what they will be like, and whether they will continue indefinitely?

This chapter takes up such questions and is structured as follows. After reviewing psychobiological and cultural explanations for human artistry, I briefly discuss patterns of trans-historical changes in the arts, noting that a rampant drive for originality, most evident in elite art forms in recent generations, has led to a postmodern 'crisis' within those traditions. Extrapolating from this present situation, I then speculate on future developments in the arts. In a nutshell, I shall argue that human nature will continue to strongly constrain the arts, especially in their more popular forms; however, when technology arises that fundamentally usurps or supplants human nature, all bets are off about what the arts will be like, what functions they may serve, and whether something akin to the arts even need endure.

7.1 The Arts and Human Nature: Culture Versus Psychobiology

Traditionally, two opposing categories of explanations have been offered for human artistry: culture and psychobiology. In practice, these forces heavily interact, but one may start the discussion by caricaturing them thus. The cultural view is exogenous: artistry is mainly propelled by social factors external to the individual artist, who is regarded as psychologically a blank slate. The particulars of a culture determine the local style, and since there is no internal counterweight to these highly variable external forces, human artistry and creativity are essentially unconstrained. Thus, the cultural view emphasizes the diversity of artistic productions; any commonalities across traditions are coincidences. A culture-focused view tends to be espoused by scholars in 'softer' domains, including postmodern theorists (Barthes 1968/2001), sociologists (Bourdieu 1979/1984; Davis 1992), art historians (Danto 1964; Elkins 2002; Pastoureau 2001; Shiner 2001), anthropologists (Bohannan 1966; Boas 1927/1955), and some psychologists (Lubart 2010; Nisbett 2003; Sawyer 2006).

In contrast, the psychobiological view is endogenous: artistry is driven by factors internal to the individual artist, whose mentality has been highly structured by evolution – as has the artist's audience. Here, cultural factors may tweak a few parameters of expression, but most artistic productions obey strongly canalized biological constraints. A psychobiological focus emphasizes what different artistic traditions have in common and seeks to identify aesthetic universals or regularities, which transcend specific cultures. This view is usually advocated by psychologists and allied social science researchers (Berlyne 1971; Keyser 2020; Martindale 2007; Pinker 2002), biologists and neuroscientists (Chatterjee and Vartanian 2014; Ishizu and Zeki 2011; Purves 2017; Wilson 1998), and some philosophers (Dutton 2009; Seeley 2019).

In their pure forms, these views cannot easily be reconciled. Each can be justified, and a forced choice between them is a matter of personal taste. Elsewhere (Kozbelt 2016, 2017a, b, 2019a, b, 2020) I have unpacked these two categories in considerable detail. On balance, I favor the psychobiological rationale for endogenous aesthetic regularities. I believe that the empirical evidence supporting that view is stronger than the cultural case for unconstrained aesthetic diversity. I also believe that starting with an assumption of evolved aesthetic regularities is a sounder research orientation – if one does not look for such regularities, one is unlikely to find them – that has more potential for understanding human nature.

Two lines of research directly support the psychobiological view: studies of aesthetic preference or judgment, and commonly observed regularities in the features of aesthetic products. First, if human aesthetics were truly a free-for-all, little general agreement should be evident in aesthetic liking and evaluation; however, many studies have found cross-cultural concordances, especially among experts (Chen et al. 2002; Dutton 2009; Ford et al. 1966). Second, rule-based regularities in aesthetic products have been amply documented across domains. For instance, visual artworks typically incorporate many aspects of color and shape preference (for summaries, see Kozbelt 2020; Martindale 2007), a center compositional balancing point (Firstov et al. 2007), scale-invariant distribution of spatial frequencies (Graham and Redies 2010), and use of the Golden ratio (Green 1995). Pervasive features of music world-wide include common pitch intervals, tonal hierarchies, principles of grouping and meter, and aspects of melodic contour (Huron 2006; Justus and Hutsler 2005; Purves 2017; Savage et al. 2015; Trehub 2000). Language-based arts feature many regularities in poetic meter and rhyme (see Keyser 2020) as well as prototypical narrative elements like the so-called Hero's journey (Campbell 1949) and Polti's (1916/1921) catalogue of basic dramatic situations.

In sum, the psychobiological view argues that human aesthetics are not arbitrary or infinitely malleable. Our capacity for aesthetic experience and the particulars of our preferences are bounded, as a legacy of remorseless evolutionary processes, which have constrained human aesthetics and creativity in enduring ways.[1] Importantly, a commitment to psychobiological explanations need not entail a one-size-fits-all aesthetic formula, nor preclude the possibility of dynamism or change. Indeed, adopting an evolutionary perspective implies some capacity for transformation, since plasticity and variation are more characteristic of functional biological organization than are rigidity and invariance (Laland et al. 2015; Marcos and Chatterjee 2019). Acknowledging some flexibility does not fatally undermine the psychobiological position, since that view does firmly insist on pervasive canalized aesthetic biases that act as default preferences or as guardrails delimiting the ultimate scope of aesthetic variability.

7.2 Insights into Human Nature from Art History and the Current 'Crisis' of the Arts

While aesthetic flexibility can blossom into different modes over time, regularities in the historical development of artistic styles themselves further inform the psychobiological and cultural facets of human nature (see Kozbelt 2021). In brief, style trends can be parsimoniously understood via psychobiological processes deployed by individual creators to achieve their aesthetic goals (Kozbelt 2017a). Such processes undergird the most thorough and well-supported scientific account of trans-historical style changes in the arts (Martindale 1990). In this model, creators compete for critical attention by producing work that is evermore attention-grabbing (higher in 'arousal potential') and original, either via 'primordial cognition' (daydream-like thought), or by developing a new style altogether. Within a tradition, arousal potential always increases over time, as creators try to outdo their predecessors and contemporaries. Primordial cognition and stylistic change simultaneously increase overall and oscillate inversely with each other as they ascend – since only one or the other method of introducing novelty is necessary to further increase arousal potential, and 'too much' novelty is counterproductive.

Martindale argued that as styles change, disorder and unpredictability always increase. Thus, aesthetic entropy will inexorably exhaust from within any aesthetic tradition worthy of the name. The endgame of this dynamic is the inability of the fine arts in their late stages to meaningfully communicate to audiences. On this note, Martindale (2009) described art's tragic end, claiming that the high arts are already

[1] A related issue is the role of the arts in having actively steered the direction of human evolution. Are the arts woven *so* deeply into our nature – through Darwinian mechanisms, say – that we could never evolve out of them, without becoming something other than human? For now, I leave this speculation hanging but will return to it near the end of the chapter.

extinct (see Murray 2003, for a similarly pessimistic take). In contrast, Keyser (2020) offered a less cynical appraisal, arguing that modernist creators simply abandoned traditional rule systems based on natural aesthetics and invented new systems, often based on private codes with hidden meanings – some of which, however, represent genuine discoveries of new aesthetic principles, like recursion in the poetry of Wallace Stevens or the deployment of fractals in Jackson Pollock's drip paintings.

In any case, developments over the past century have raised many challenges to our traditional understanding of the arts. The uncertain condition of the contemporary arts is often labeled 'postmodern.' That term means different things to different people – I suppose that is the point – but functionally it boils down to a pluralistic milieu characterized by a lack of agreed-upon meaning, heterogeneous styles, short-lived artistic movements, idiosyncratic and one-trick creators making desperate ploys for originality, and a suspension of value judgments. Elkins (2002) explored the long-term prospects for such a cultural situation, in turn considering postmodernism a transient stylistic period like any other, or an intractable, indefinitely lingering endgame condition that is unclear about whether the 'game' of art can ever be ended, or some new kind of game that commences in the aftermath of art.

It is hard to know what to make of these possibilities, or what recent developments in the arts mean from an evolutionary or psychobiological perspective. Does postmodern art represent a final triumph of socio-cultural forces over biology, liberating us from the constraints of our evolutionary past? Or are the contemporary arts still evolutionarily constrained, just perhaps in ways that we do not yet understand? And are the prospects for the arts genuinely bleak, *à la* Martindale (2009) or Elkins (2002): a forced choice between death or perpetual postmodern limbo?

7.3 The Artwork of the Future?

There are numerous reasons to think that the arts still have something of a future. For instance, one may question some of Martindale's assumptions, which would undermine the logical necessity of his dark conclusion (see Kozbelt 2009a, 2017a, 2019b). One important distinction is that the psychobiological mechanisms for *capturing* attention – so prominent in Berlyne (1971) and Martindale (1990) – may have been erroneously extrapolated to the process of *sustaining* critical attention over the generations. Another objectionable point involves Martindale's (2009) baffling remarks about modern visual art comprising random marks on canvases, thus having no variability, and thus not being subject to further evolutionary change. This is at odds with Elkins's (2002) view that contemporary artists feel pressure to develop instantly recognizable styles, as well as empirical research documenting viewers' impressive discrimination ability, even among non-experts and even for highly abstract paintings (Hawley-Dolan and Winner 2011).

Another point of caution is that art historical systems are messy and contingent. Even the pervasive art historical trends discovered by Martindale (1990) do not imply a deterministically ratcheted progression. There are many ways to be original.

There is no linear ordering of novelty in aesthetic innovations, which would preordain the specific direction of a tradition, and art historical systems do not have a perfect memory or perfectly calibrated standards of innovation anyway. If they did, a single instance of extreme novelty would preclude later but less radical works from ever garnering attention, and no music at all would have been written after John Cage's silent *4 minutes and 33 seconds*. Rather than a neat sequence, a better paradigm is that aesthetic styles inhabit a bushy set of niches, which coexist and interact in complex ways, and which emerge in socio-cultural settings through the personalities of individual creators.

Forecasting a less bleak future for the arts implies long-term sustainability and flexibility. This can take several forms (Kozbelt 2019b, 2020). One theoretical option is that cultural forces are more important than I have suggested, despite the evidence cited above. That is, perhaps alleged aesthetic universals and observed patterns in the history of the arts are illusory and represent the easiest options for structuring aesthetic behavior, but with no inherent constraints on future artistry. Or perhaps cultural forces, including technology, initially build on an evolutionary foundation but continually gain in importance over historical time, eventually diminishing meaningful psychobiological constraints on aesthetics.

Splitting the difference between an account where human aesthetic biases are inviolable and the fine arts have already died versus one where a blank-slate aesthetic allows culture to completely trump biology, a sensible middle-ground option is the long-term co-evolution of aesthetic regularities among creators and audiences (Kozbelt 2020). In this view, performances or artifacts and their evaluations mutually transform each other through their history of interactions, and "art consists of a form of communication that has coevolved with its evaluation" (Prum 2013, p. 818). This is consistent with the psychobiological and historical evidence presented above and reflects an inherent plasticity: while psychobiological constraints on human aesthetics and creativity have a long evolutionary pedigree and cannot be excised from human nature, they are not completely fixed for all time.

Indeed, as creators forge new styles and modes of art, they sometimes hit upon novel principles that represent true aesthetic discoveries, rather than mere inventions. Thus, previously unrealized aesthetic regularities may become evident in historical times.[2] Examples may include poetic recursion and painted fractals (Keyser 2020), a center compositional balancing point in painting (Firstov et al. 2007), linear perspective in visual art (Kubovy 1988), featural commonalities across different orthographic systems (Gleitman and Rozin 1977; Treiman and Kessler 2011), and a common 'language of film' (Edgar et al. 2015). New aesthetic principles can promote artistic variability, while maintaining links to the evolved structure of the human mind. Over time, different innovations may also be subject to higher-order recombination and synthesis – as in a Wagnerian *Gesamtkunstwerk* or modern cinema (Brown and Dissanayake 2018) – with their own emergent properties

[2] A salutary point, since evolutionary aesthetics and the creative process are usually discussed in largely ahistorical terms (see also Kozbelt 2009b, 2020).

providing grist for further development. Indeed, understanding the enduring quality of some new aesthetic principles – in contrast to transient aesthetic fads – seems a rich area for future research, both in quantitatively documenting such instances and in trying to understand the basis of their 'adaptive' worth and staying power (see Gombrich 1979).

Yet another mode of promoting ongoing aesthetic sustainability involves a question of values, in rethinking the standard criteria of creativity – originality, value, and surprise (Simonton 2012). Most accounts of the fine arts in the West, including Martindale (1990, 2009), prize originality above all else. However, not all modes of creativity work that way. For instance, Galenson (2001, 2006, this volume) has documented the remarkable ability of some creators (dubbed 'seekers,' as opposed to 'finders') to avoid creative burnout and continually improve the quality of their work as they age. They do this by prioritizing elaborative processes, technical skill, expertise, sound judgment, flexibility in the creative process, and the virtues of building on tradition – at the expense of raw hit-or-miss ideation, radical originality for its own sake, and rejecting tradition (see also Kozbelt 2009b, c, 2019b). Seekers' mode of operation has yielded some of the finest aesthetic accomplishments in history, but the greatness of their contributions is harder to articulate, and they tend to receive less critical attention than finders (Gombrich 1991; Kozbelt 2014).

The potential for sustainable culture-level aesthetic and creative learning in the mode of seekers represents one of the most provocative dimensions of Galenson's theory. Overall, there are many lessons about long-term creative sustainability that can be derived from the values and activities of artistic seekers (see Kozbelt and Durmysheva 2007), which can counter the culturally pernicious effects of fetishizing originality. Particularly in a postmodern context, it is unclear how such a cultural reorientation might be achieved. But drawing attention to this possibility – by emphasizing the virtues of wisdom, purpose, individual and cultural learning, transcendental good, and sustainably building on tradition – seems like a prudent and necessary first step.

7.4 The Popular Arts and Human Nature Revisited

More popular modes of art represent another potential path. In contrast to the moribund high arts, popular arts are alive and well and – at their best – probably better than they have ever been (Pinker 2002). Popular forms of art, by definition, speak directly to salient aspects of human nature and are less mediated by elitism (Bourdieu 1979/1984; Pinker 1997), even though nowadays they are subject to sophisticated critical evaluation.

Much theoretical hand-wringing about the current unprecedented state of the arts is based on a categorical dichotomy between elite versus popular art (Martindale 2009; Murray 2003) – with the implication that the distinction is starker now than in the past. However, the historical relation between the two is difficult to evaluate quantitatively, since the preserved cultural record is biased toward high art, especially in more distant epochs. Moreover, the categories of the high arts and popular arts are themselves slippery. Thus, many basic questions remain unanswered. For instance, assuming popular arts were the original mode of human aesthetics, when and how did high art traditions emerge? Has the relation between the two systematically changed over time? What cultural conditions favor stronger versus weaker distinctions between the high and popular arts? Have the high arts always been elitist, in the way they are today? Is it possible that previously the high arts were, relative to the population, *more* elitist than today (Csikszentmihalyi 1988)? What explains blending or exchange between the two, or the cross-over appeal of authentically difficult artists like Shakespeare and Victor Hugo, Mozart and Beethoven, or Michelangelo and Monet, to the popular imagination?

For good measure, an additional wrinkle concerns the nature of subjective aesthetic experience – surely a key reason for the arts' enduring appeal. How does aesthetic experience vary across the fine arts-popular arts spectrum? Are aesthetic emotions similar in quality or intensity across that spectrum, and how might aesthetic emotions differ from ordinary emotions (Konečni 2013)? How does an audience's level of knowledge, skill, or expertise impact aesthetic experience (Bullot and Reber 2013)? How *de rigueur* are specialized armaments of historical, biographical, or technical knowledge for different kinds of aesthetic experiences? Have ivory-tower modes of theoretical or critical analysis exacerbated the current crisis of the high arts (Elkins 1998; Pinker 2002)?

Given all these uncertainties, perhaps it is not insane to argue that the broad-brush picture of the arts today is not so wildly different from in the past. We have long had a complex cultural aesthetic system consisting of many niches with different characteristics and dynamics and audiences and selection pressures, but numerically dominant popular forms partaking of humanistic natural aesthetic principles, with hedonic enjoyment of beauty and meaning-making paramount in aesthetic experience. If the contemporary high arts have made themselves irrelevant for most people, this just reinforces the enduring power of long-standing psychobiological principles within a humanistic conception of the arts.[3] Being rooted in human nature, the popular arts seem unlikely to change radically in the foreseeable future, staying mostly within the guardrails of the evolved structure of the mind, despite changes of media and forms and styles, and perhaps slowly evolving in a give-and-take process whereby the raw material of earlier aesthetic efforts is continually recycled in new ways. Thus, even as the arts appear to change, our basic human nature does not.

Indeed, after all these millennia, what could possibly change *that*?

---

[3] As an unrepentant aesthetic elitist, I note that much contemporary popular and high art is terrible, that even small audiences may suffice for the continuation of elite traditions, and that in the long run even super-niche works may have considerable staying power in the culture – assuming they partake of, rather than utterly reject, 'natural' aesthetic rules in line with evolved psychobiological principles.

7.5 The Wildcard of Technology

The cultural force with the greatest capacity to induce basic changes to our humanity is technology. This is not a new phenomenon: changes in stone tool technology have accompanied our species' evolutionary emergence over the last few million years, and in historical times new technologies have enabled dramatic expansions of cultural practices (Kozbelt 2019a). Technology has likewise long played a major role in the development of the arts (e.g., Ball 2003; Gombrich 1991; Pastoureau 2001). It is easy to forget how difficult and expensive it was to manufacture durable pigments or musical instruments in the pre-modern era, how arcane even basic artistic technical knowledge was, and how little access most people had to aesthetically significant works until very recently. The historical emergence of domains like photography and cinema underscores the potential for new technology to act as a midwife for artistic expression.

Today, computer technology and the internet have facilitated an unprecedented production and dissemination of information in general, and this applies to the arts as well. The finest works of literature, painting, and music are immediately accessible online, and aspiring artists can easily use the internet to improve their craft – teenage musicians need no longer take a bus or two across Liverpool just to learn a B^7 guitar chord. While most of the visual art, music, and literary contributions circulated online resemble what one would encounter in real life, recent technology has also fostered domain-specific aesthetic innovations. These include new building materials in architecture, hypertext possibilities in literary arts, synthesized timbres and sound-manipulation methods in music, computer-generated imagery in visual art and film, and technology-dependent installation art like Olafur Eliasson's *Weather Project* at the Tate Modern ("Weather Project" 2003) and Random International's *Rain Room* at MoMA ("Rain Room" 2013). The use of technology to guide the creative process is also pervasive, including recording technology for musical or dance improvisations, word processing functionality in writing, and computer-aided design in art and architecture. Digital technology has also made possible new domains with aesthetic potential, including virtual and augmented reality, video games, sound generative sculpture, and other emerging hybridized art forms. And while "cultural prophecy is always a mug's game" (Bloom 1994, p. 548), stimulating and rampantly varied speculations about the future intersection of the arts and technology are easy to find online (e.g., "Predicting the Future," 2020; "What Art Will Look Like in 100 Years" 2020).

Do these technological developments have the potential to fundamentally change human nature? At first blush, this might seem unlikely. Throughout this chapter, I have argued that during historical times, the arts have largely followed aesthetic channels laid down in our evolutionary past. Most people have conservative artistic tastes, and even card-carrying aesthetes can be skeptical of newfangled trends: as eminent literary critic Harold Bloom mused, "Sometimes I try to visualize Dr. Johnson or George Eliot confronting MTV Rap or experiencing Virtual Reality and find myself heartened by what I believe would be their ironical, strong refusal of such irrational entertainments" (Bloom 1994, p. 517).

However, in line with psychobiological plasticity (Laland et al. 2015) and a long-term co-evolutionary dynamic (Prum 2013), perhaps some aspects of our humanity have been modified – if not completely transformed – by our experience with the arts. For instance, a supposed important, multifarious, and civilizing benefit of the arts is the claim that they promote "the enlargement of human faculties" (Clark 1969, p. 323). This might include learning new mental frames for understanding the world (Gablik 1976), enhanced perceptual abilities (Kozbelt and Seeley 2007; Moreno and Bidelman 2014), a suite of cognitive, attitudinal, behavioral, health, and economic effects (McCarthy et al. 2004), moral judgments (see Devereaux 2004), or simply an enrichment of subjective experience (Dewey 1934). Such art-based changes echo broader societal trends related to technology, like rising IQ scores in the industrialized world in recent decades (Flynn 2009) and arguments about the broadly beneficial effects of the increasing complexity of popular entertainment and media (Johnson 2005). Thus, if in recent generations the arts have changed faster than ever, perhaps audiences have also changed faster than ever. Does the mutual growth spurred by this aesthetic arms race amount only to a tweaking of cognitive parameters such that most emerging art will remain within the scope of evolved aesthetic biases, or is something bigger afoot?

The technologies discussed above are largely tools that facilitate many longstanding artistic practices, with humans still firmly in charge. More disquieting in their long-term implications are artificial intelligence (AI) programs that produce their own work, rather than merely supporting human creativity. Examples abound in many domains in the arts and sciences (Du Sautoy 2019; Miller 2020). For example, well-known autonomous visual art AI programs include Harold Cohen's innovative AARON (see McCorduck 1991) as well as more recent efforts like The Painting Fool (Colton and Wiggins 2012) and AICAN (Mazzone and Elgammal 2019), which produce novel work based on corpuses of art historical images, subject to stylistic modification. Generative music programs emulating the styles of eminent composers were pioneered by David Cope (see Cope 2006), and more recent AI work in music has a significant online presence (e.g., "AI Music Creativity," 2021; "Artificial Intelligence Composing Emotional Soundtrack Music" 2021). Such cases are the tip of the AI iceberg, which seems bound to develop apace. Opinions about the quality of machine-produced work differ, with many remaining unimpressed (see Sawyer 2006; Tang 2020). However, given Moore's law on ongoing improvements in computing power, and parallel developments in, say, the domain of chess – where the best machines outplay the best human players (see Higgins 2017) – it is possible that human aesthetics may be headed toward a so-called singularity (Kurzweil 2005), where machine intelligence becomes so high and the rate of change so fast that it becomes impossible for us to predict anything about the future, and the fusion of human and artificial intelligence radically and irrevocably transforms life (see also Harari 2017).

In considering the fate of such technologies – and ourselves, as part of the bargain – little can be said with any certainty. But to the extent there is a continued place in the world for flesh-and-blood-and-brain humans, I suspect that the arts will remain integral to our species. In meditating upon the prospect of far more advanced beings, it is tempting to imagine creative and aesthetic sensibilities commensurate with their computational powers, akin to the God of Genesis contemplating His handiwork on the seventh day. But it is also worth entertaining the possibility that such beings may find the arts, at least as we understand them, *passé*. Perhaps from the widest vantage point, aesthetic experience thrives only in a relatively narrow computational band. Simple organisms lack the processing capacity and self-awareness to engage in any aesthetic behavior. As our species evolved, artistic creativity emerged and became a signature faculty. Human aesthetic experience depends on our cognitive endowments, but it is equally rooted in our cognitive limitations: to be instantly and fully aware of the implications of an aesthetic work is to cease to have an aesthetic experience. It is conceivable that advanced beings, approaching the level of Laplace's fabled demon, would have better things to do than art. But I personally cannot imagine what those things might be.

In the meantime, I can revel in my Shakespeare, Bach, and Rembrandt, confident that they will continue to reward my efforts to uncover their secrets, and that will have to be enough.

References

"AI Music Creativity." (2021.) https://aimusiccreativity.org/. Accessed March 25, 2021.

"Artificial Intelligence composing emotional soundtrack music." (2021). https://www.aiva.ai/. Accessed March 25, 2021.

Ball, P. (2003). *Bright earth: Art and the invention of color.* Chicago: University of Chicago Press.

Barthes, R. (1968/2001). The death of the author. In V. B. Leitch (Eds.), *Norton anthology of theory and criticism* (pp. 1466-1470). New York: Norton.

Berlyne, D. E. (1971). *Aesthetics and psychobiology.* New York: Appleton-Century-Crofts.

Bloom, H. (1994). *The Western canon: The books and school of the ages.* New York: Harcourt Brace & Company.

Boas, F. (1927/1955). *Primitive art.* New York: Dover.

Bohannan, L. (1966). Shakespeare in the bush: An American anthropologist set out to study the Tiv of West Africa and was taught the true meaning of Hamlet. *Natural History, 75*, 28-33.

Bourdieu, P. (1979/1984). *Distinction: A social critique of the judgement of taste.* Cambridge, MA: Harvard University Press. (Originally published in 1979 by Les Editions de Minuit, Paris, as *La Distinction: Critique sociale du jugement.*)

Brown, S., & Dissanayake, E. (2018). The synthesis of the arts: From ceremonial ritual to "Total Work of Art." *Frontiers of Sociology, 3,* 9. doi: https://doi.org/10.3389/fsoc.2018.00009

Bullot, N. J., & Reber, R. (2013). The artful mind meets art history: Toward a psycho-historical framework for the science of art appreciation. *Behavioral and Brain Sciences, 36,* 123-137.

Campbell, J. (1949). *The hero with a thousand faces.* New York: Pantheon.

Chatterjee, A., & Vartanian, O. (2014). Neuroaesthetics. *Trends in Cognitive Sciences, 18,* 370-375.

Chen, C., Kasof, J., Himsel, A.J., Greenberger, E., Dong, Q., & Xue, G. (2002). Creativity in drawings of geometric shapes: A cross-cultural examination with the Consensual Assessment Technique. *Journal of Cross-Cultural Psychology, 33,* 171-187.

Clark, K. (1969). *Civilisation: A personal view.* New York: Harper & Row.

Colton, S., & Wiggins, G. A. (2012). Computational creativity: The final frontier? In L. De Raedt, C. Bessiere, D. Dubois, P. Doherty, & P. Frasconi (Eds.), *ECAI'12: Proceedings of the 20th European Conference on Artificial Intelligence* (pp. 21-26). Amsterdam: IOS Press.
Cope, D. (2006). *Computer models of musical creativity*. Cambridge, MA: MIT Press.
Csikszentmihalyi, M. (1988). Society, culture, and person: A systems view of creativity. In R. J. Sternberg (Ed.), *The nature of creativity: Contemporary psychological perspectives* (pp. 325–228). New York: Cambridge University Press.
Danto, A. (1964). The Artworld. *Journal of Philosophy, 61,* 571-584.
Davis, F. (1992). *Fashion, culture, and identity*. Chicago: University of Chicago Press.
Devereaux, M. (2004). Moral judgments and works of art: The case of narrative literature. *Journal of Aesthetics and Art Criticism, 62,* 3-11.
Dewey, J. (1934). *Art and experience*. New York: Capricorn Books.
Du Sautoy, M. (2019). *The creativity code: Art and innovation in the age of AI*. Cambridge, MA: Harvard University Press.
Dutton, D. (2009). *The art instinct: Beauty, pleasure, and human evolution*. New York: Bloomsbury Press.
Edgar, R., Marland, J., & Rawle, S. (Eds.) (2015). *The language of film*. New York: Bloomsbury Publishing.
Elkins, J. (1998). *Why are our pictures puzzles? On the modern origins of pictorial complexity*. New York: Routledge.
Elkins, J. (2002). *Stories of art*. New York: Routledge.
Firstov, V., Firstov, V., Voloshinov, A., & Locher, P. (2007). The colorimetric barycenter of paintings. *Empirical Studies of the Arts, 25,* 209-217.
Flynn, J. R. (2009). *What is intelligence? Beyond the Flynn effect*. New York: Cambridge University Press.
Ford, C. S., Prothro, E. T., & Child, I. L. (1966). Some transcultural comparisons of esthetic judgment. *The Journal of Social Psychology, 68,* 19-26.
Gablik, S. (1976). *Progress in art*. New York: Rizzoli.
Galenson, D. W. (2001). *Painting outside the lines: Patterns of creativity in modern art*. Cambridge, MA: Harvard University Press.
Galenson, D. W. (2006). *Old masters and young geniuses: The two life cycles of artistic creativity*. Princeton, NJ: Princeton University Press.
Gleitman, L. R., & Rozin, P. (1977). The structure and acquisition of reading I: Relations between orthographies and the structure of language. In A. S. Reber & D. L. Scarborough (Eds.), *Toward a psychology of reading: The proceedings of the CUNY* conferences (pp. 1-53). Hillsdale, NJ: Lawrence Erlbaum Associates.
Gombrich, E. H. (1979). *The sense of order: A study in the psychology of decorative art*. London: Phaidon.
Gombrich, E. H. (1991). *Topics of our time: Twentieth-century issues in learning and in art*. London: Phaidon.
Graham, D. J., & Redies, C. (2010). Statistical regularities in art: Relations with visual coding and perception. *Vision Research, 50,* 1503-1509.
Green, C. D. (1995). All that glitters: A review of psychological research on the aesthetics of the golden section. *Perception, 24,* 937-968.
Harari, Y. N. (2017). *Homo deus: A brief history of tomorrow*. New York: Harper Perennial.
Hawley-Dolan, A. & Winner, E. (2011). Seeing the mind behind the art: People can distinguish abstract expressionist paintings from highly similar paintings by children, chimps, monkeys, and elephants. *Psychological Science, 22,* 435-441.
Higgins, C. (2017). "A brief history of Deep Blue, IBM's chess computer." https://www.mentalfloss.com/article/503178/brief-history-deep-blue-ibms-chess-computer. Accessed March 25, 2021.
Huron, D. (2006). *Sweet anticipation: Music and the psychology of expectation*. Cambridge, MA: MIT Press.

Ishizu, T., & Zeki, S. (2011). Toward a brain-based theory of beauty. *PLoS One, 7,* e21852.
Johnson, S. (2005). *Everything bad is good for you: How today's popular culture is actually making us smarter.* New York: Riverhead Books.
Justus, T., & Hutsler, J. J. (2005). Fundamental issues in the evolutionary psychology of music: Assessing innateness and domain specificity. *Music Perception, 23,* 1-27.
Keyser, S. J. (2020). *The mental life of modernism: Why poetry, painting, and music changed at the turn of the twentieth century.* Cambridge, MA: MIT Press.
Konečni, V. (2013). A critique of emotivism in aesthetic accounts of visual art. *Philosophy Today, 57,* 388-400.
Kozbelt, A. (2009a). Maybe, maybe even probably, but not necessarily: Response to Martindale. *Empirical Studies of the Arts, 27,* 167-171.
Kozbelt, A. (2009b). The evolution of evolvability, applied to human creativity. *International Journal of Creativity and Problem Solving, 19,* 101-121.
Kozbelt, A. (2009c). Ontogenetic heterochrony and the creative process in visual art: A précis. *Psychology of Aesthetics, Creativity, and the Arts, 3,* 35-37.
Kozbelt, A. (2014). Musical creativity across the lifespan. In D.K. Simonton (Ed.), *Handbook of genius* (pp. 451-472). Hoboken, NJ: Wiley-Blackwell.
Kozbelt, A. (2016). Creativity and culture in visual art. In V. Glăveanu (Ed.), *The Palgrave handbook of creativity and culture research* (pp. 573-594). New York: Palgrave Macmillan.
Kozbelt, A. (2017a). Tensions in naturalistic, evolutionary explanations of aesthetic reception and production. *New Ideas in Psychology, 47,* 113-120.
Kozbelt, A. (2017b). Musical creativity. In J. C. Kaufman, V. Glăveanu, & J. Baer (Eds.), *The Cambridge handbook of creativity across domains* (pp. 161-180). New York: Cambridge University Press.
Kozbelt, A. (2019a). Evolutionary approaches to creativity. In J. C. Kaufman & R. J. Sternberg (Eds.), *The Cambridge handbook of creativity, 2nd ed.* (pp. 109-131). New York: Cambridge University Press.
Kozbelt, A. (2019b). Evolutionary constraints on creativity. In L. Dorfman, P. Machotka, & V. Petrov (Eds.), *Integrative explorations of the creative mind* (pp. 79-100). Newcastle upon Tyne, UK: Cambridge Scholars Publishing.
Kozbelt, A. (2020). Evolutionary constraints on the visual and plastic arts. In J. Carroll, M. Clasen, & E. Jonsson (Eds.), *Evolutionary perspectives on imaginative culture* (pp. 213-232). New York: Springer.
Kozbelt, A. (2021). The history of the arts as a window into human nature. *Frontiers in Psychology, 12,* 787238. https://doi.org/10.3389/fpsyg.2021.787238
Kozbelt, A., & Durmysheva, Y. (2007). Lifespan creativity in a non-Western artistic tradition: A study of Japanese ukiyo-e printmakers. *International Journal of Aging and Human Development, 65,* 23-51.
Kozbelt, A., & Seeley, W. P. (2007). Integrating art historical, psychological, and neuroscientific explanations of artists' advantages in drawing and perception. *Psychology of Aesthetics, Creativity, and the Arts, 1,* 80-90.
Kubovy, M. (1988). *The psychology of perspective and Renaissance art.* New York: Cambridge University Press.
Kurzweil, R. (2005). *The singularity is near: When humans transcend biology.* New York: Viking.
Laland, K. N., Uller, T., Feldman, M. W., Sterelny, K., Müller, G. B., Moczek, A., Jablonka, E., & Odling-Smee, J. F. (2015). The extended evolutionary synthesis: Its structure, assumptions and predictions. *Proceedings of the Royal Society B: Biological Sciences, 282,* 20151019.
Lubart, T. I. (2010). Cross-cultural perspectives on creativity. In J. C. Kaufman & R. J. Sternberg (Eds.), *Th e Cambridge handbook of creativity* (pp. 265–278). New York: Cambridge University Pres.

Marcos, N., & Chatterjee, A. (2019). Neuroaesthetics and art's diversity and universality. *WIREs Cognitive Science, 10*(3). https://doi.org/10.1002/wcs.1487

Martindale, C. (1990). *The clockwork muse: The predictability of artistic change*. New York: Basic Books.

Martindale, C. (2007). A neural-network theory of beauty. In C. Martindale, P. Locher, & V. M. Petrov (Eds.), *Evolutionary and neurocognitive approaches to aesthetics, creativity, and the arts* (pp. 181-194). Amityville, NY: Baywood.

Martindale, C. (2009). The evolution and end of art as Hegelian tragedy. *Empirical Studies of the Arts, 27,* 133-140.

Mazzone, M., & Elgammal, A. (2019). Art, creativity, and the potential of artificial intelligence. *Arts, 8*, 26.

McCarthy, K. F., Ondaatje, E. H., Zakaras, L., & Brooks, A. (2004). *Gifts of the muse: Reframing the debate about the benefits of the arts*. Santa Monica, CA: The RAND Corporation.

McCorduck, P. (1991). *AARON's code: Meta-art, artificial intelligence, and the work of Harold Cohen*. New York: W.H. Freeman.

Miller, A. I. (2020). *The artist in the machine: The world of AI-powered creativity*. Cambridge, MA: MIT Press.

Moreno, S., & Bidelman, G. M. (2014). Examining neural plasticity and cognitive benefit through the unique lens of musical training. *Hearing Research, 308*, 84-97.

Murray, C. (2003). *Human accomplishment: The pursuit of excellence in the arts and sciences*, 800 B.C. to 1950. New York: HarperCollins.

Nisbett, R. (2003). *The geography of thought: How Asians and Westerners think differently...and why*. New York: Free Press.

Pastoureau, M. (2001). *Blue: The history of a color*. (Trans. M. I. Cruse). Princeton, NJ: Princeton University Press.

Pinker, S. (1997). *How the mind works*. New York: Norton.

Pinker, S. (2002). *The blank slate: The modern denial of human nature*. New York: Penguin.

Polti, G. (1916/1921). *The thirty-six dramatic situations*. Franklin, OH: James Knapp Reeve.

Prum, R. O. (2013). Coevolutionary aesthetics in human and biotic artworlds. *Biology & Philosophy, 28*, 811-832. https://doi.org/10.1007/s10539-013-9389-8

"Predicting the Future." (2020). https://contemporarylynx.co.uk/predicting-the-future-visual-arts-in-2050. Accessed March 23, 2021.

Purves, D. (2017). *Music as biology: The tones we like and why*. Cambridge, MA: Harvard University Press.

"Rain Room." (2013). https://www.moma.org/calendar/exhibitions/1352. Accessed March 25, 2021.

Savage, P. E., Brown, S., Sakai, E., & Currie, T. E. (2015). Statistical universals reveal the structures and functions of human music. *Proceedings of the National Academy of Sciences of the United States of America, 112*, 8987-8992.

Sawyer, R. K. (2006). *Explaining creativity: The science of human innovation*. New York: Oxford University Press.

Seeley, W. P. (2019). *Attentional engines: A perceptual theory of the arts*. New York: Oxford University Press.

Shiner, L. (2001). *The invention of art: A cultural history*. Chicago: University of Chicago Press.

Simonton, D. K. (2012). Taking the U.S. Patent Office criteria seriously: A quantitative three-criterion creativity definition and its implications. *Creativity Research Journal, 24,* 97-106.

Tang, D. (2020) "The machines are coming, and they write really bad poetry." https://lithub.com/the-machines-are-coming-and-they-write-really-bad-poetry/. Accessed March 25, 2021.

Trehub, S. E. (2000). Human processing predispositions and musical universals. In N. L. Wallin, B. Merker, & S. Brown (Eds.), *The origins of music* (pp. 427-448). Cambridge, MA: MIT Press.

Treiman, R., & Kessler, B. (2011). Similarities among the shapes of writing and their effects on learning. *Written Language and Literacy, 14*, 39–57.
"Weather Project" (2003). https://www.tate.org.uk/whats-on/tate-modern/exhibition/unilever-series/unilever-series-olafur-eliasson-weather-project. Accessed March 25, 2021.
"What Art Will Look Like in 100 Years." (2020). http://momus.ca/what-will-art-look-like-in-100-years-sixteen-contemporary-artists-predict-the-future/. Accessed March 23, 2021.
Wilson, E. O. (1998). *Consilience: The unity of knowledge.* New York: Vintage.

Aaron Kozbelt (PhD, The University of Chicago, 2002) is a Professor and Chairperson of the Psychology Department at Brooklyn College and is also affiliated with the Graduate Center of the City University of New York. His research focuses on creativity and cognition in the arts. He has published more than eighty peer-reviewed articles and book chapters and is a co-editor of the 2018 Cambridge Handbook of Expertise and Expert Performance (2nd ed.). He serves on several editorial boards and has received several national and international awards for his research.

Chapter 8
Embodied Creativity and Technology: A Complex Relationship

Paula Thomson and S. Victoria Jaque

Abstract Cognitive/creative embodiment is a memory system that encodes information about physical competencies, contextual perceptions, and motor responses to internal and external situations. Elite athletes and performers cultivate highly refined embodied awareness that directly influences creative expression. Embodied creativity draws upon sensorimotor, relational, emotional, and aesthetic perceptions and actions. Technology is a powerful tool that is integrated into the training and promotion of these populations. Embodied creativity and embodied cognition are dependent on real-time physical engagement; however, physical engagement continues to rapidly change in response to technological advancements. During the recent COVID 19 pandemic, the interface of technology and elite performance has revealed unique possibilities as well as profound limitations. Embodiment, creativity and their interaction with technology are complex and dynamic processes. In order to navigate the changes that result from these interacting components (embodiment, learning, creativity, technology) strong adaptive skills are necessary; these challenges will equally require a strong embodied and creative self that can traverse a complex world.

Keywords Creativity · Embodiment · Pandemic · Performing arts · Sport · Technology

8.1 Introduction

Creativity, technology, and education are powerfully interconnected; investigations into improving these three factors are discussed and examined with great regularity (U. S. Department of Education 2017). The concept of embodiment, although less researched, captures the need to expand the binary limits of the mind/body and self/other (Malinin 2019). Similarly, efforts to examine creativity as a network process

P. Thomson (✉) · S. V. Jaque
California State University, Northridge, CA, USA
e-mail: paula.thomson@csun.edu

D. Henriksen, P. Mishra (eds.), *Creative Provocations: Speculations on the Future of Creativity, Technology & Learning*, Creativity Theory and Action in Education 7,
https://doi.org/10.1007/978-3-031-14549-0_8

push the limited boundaries between physical and digital spaces and the collaborative interactions among these seemingly different spaces (Gaggioli et al. 2020). Creative potential is multifaceted and ideally should include the sociocultural, real-time action, and virtual worlds. These potentials can operate at the small c-creativity (everyday) level just as much as the Big-C creativity (pre-eminent creators) level (Corazza and Glăveanu 2020).

In this chapter, embodiment is examined as a concept and is applied to the multi-layered fields of creativity, learning, and technology. By specifically including research work on two elite populations, athletes and performing artists, it is hoped that a more refined appreciation about the process of embodiment will be revealed. Technological advances are interconnected with physical and performance outcomes; they are shaping the training and delivery of sport and performing arts. Strengths and limitations associated with embodiment, creativity, and technology are discussed, including a brief exploration related to the effects of social isolation during a pandemic. The primary purpose of this chapter is to reinforce the need to integrate an understanding about embodiment within education, creativity, and technology, especially since technological advances accelerate interactions within an ever-expanding virtual world.

In order to clarify the authors' conceptualization regarding creativity, technology and learning, a few guiding principles are explicitly outlined. First, creativity can be examined at the product or process level. Our focus remains at the process level regardless of how a product might be evaluated. Factors such as absorption, frustration tolerance, persistence, problem-solving skills, preparation, goal clarity, anxiety, and a sense of power and pleasure are features of distinct creative experiences. These components overlap with optimal learning environments. Engaging with a problem and exploring ways to improve performance (goal-setting, concentration, sense of control, acquiring skills to meet increasing challenges, and unambiguous feedback during the learning and performance process) are vital within sport and the performing arts. Lastly, technology in this chapter is examined as an interactive tool that promotes opportunities to improve skills. Technology can facilitate learning for able and disabled individuals and can augment learning environments.

8.2 Embodiment

Concepts of embodiment are not new; however, recent theoretical research has shifted the focus and relevance of embodiment. Studies that investigate the complex interconnections of sensorimotor and cognitive processing are amplifying the importance of embodiment (Price et al. 2009). Fields as diverse as philosophy, psychology, cognitive science, neuroscience, sociology and technology are converging to create a dynamic, inclusive concept of embodiment (Farr et al. 2012; Malinin 2019). Currently, embodiment implies a memory system that encodes information about internal and external physiologic, emotional, and cognitive responses. Embodied cognition operates pre-reflectively; movement responses are mobilized with incredible precision and accuracy (Rietveld 2008). It is part of the implicit memory system guiding knowledge and understanding about movement, intention, and perceived emotional states in others (Dietrich 2007, 2008; Grafton 2009). Embodied states incorporate neurological networks responsible for actions and perceptions, physiologic autonomic responses, and conceptual systems such as creative aesthetics, empathy, moral values, meaning-making, and memories and imaginings of past, present and future situations (Critchley 2009; Koziol et al. 2012; Lakoff 2012; Montero 2006; Perlovsky 2015). The concept of embodiment implies an intrinsic connection between being moved by movement at the perceptual impressionistic level and being moved to move at the action and expressive level. This reiterative process of being moved by movement and being moved to move influences emotional responses; sensorimotor and emotional processes are dynamic and entwined (Fuchs and Koch 2014). Embodied interoceptive representations also shape higher cognitive processing such as symbol formation and language (Herbert and Pollatos 2012); collectively these interconnecting neural networks give life to an embodied self (Jackson and Decety 2004).

Ultimately, embodied cognition influences all psychosocial, cognitive, and creative behaviors (Price et al. 2009). Throughout individual development these embodied cognitions mature; however, even during prenatal development they are already influencing all internal and external responses and simultaneously being modified by these interactions (Llinas 2001). Our sense of embodiment continually adapts to changes throughout the lifespan. The specific body we inhabit, the physical training we receive, and the context in which we exist all inform and shape moment-to-moment embodied cognitions, perceptions, beliefs, and behaviors (Borghi and Cimatti 2010; Sze et al. 2010). An embodied self-forms from lived-experiences and an autobiographical self emerges (Llinas 2001). Regardless of whether these experiences are positive or negative, body memories are stored in an embodied representation of the self (Critchley 2009). For example, children who exhibit early signs of talent in a specific sport or arts activity will form deeply etched memories of their first time witnessing a sporting event or performance. Persistent play behavior associated with their talent is observed in these children (Olszewski-Kubilius et al. 2015). These events become part of an embodied self.

8.3 Embodiment in Elite Performers

In populations such as elite athletes and performing artists, the subjective experiences of embodiment have layers of meaning, including when an embodied sense of self is challenged during injury, illness, or extreme stress. To attain elite status in these fields, years of dedicated deliberate practice must occur (Ericsson 2013; Ericsson et al. 2007). Skill acquisition is achieved when a confluence of biomechanical factors is integrated. Basic biomechanical components include (Oglesby et al. 2018):

- Strength (ability to exert force)
- Agility (ability to apply explosive movements to rapidly change directions)
- Speed (rate of motion or rate of change in a position; includes reaction time or quickness)
- Accuracy (ability to control movement in a given direction and intensity)
- Endurance (ability to endure via cardiorespiratory, muscular, speed, or strength; implies stamina)
- Power (rate at which work is performed and includes strength and speed)
- Coordination (ability to integrate all components of physical fitness and skills; involves rhythm, spatial orientation, reaction to external and internal stimuli; involves gross motor and fine motor activation)
- Flexibility (range of motion in a joint or group of joints; includes both active and passive range of motion)
- Balance (ability to sustain precise control over the body's positions and movements)

At elite levels, years of participating in rigorous training programs address these functional biomechanical principles. This is accomplished via engagement in persistent deliberate training, which facilitates relatively permanent changes in motor skill performance.

Beyond the application of biomechanical principles listed above, motor learning strategies must be cultivated and refined. The basic motor learning strategies include three phases: (1) cognitive (gathering information), (2) associative (integrating actions), and (3) autonomous (shifting the performance of skill to a non-conscious procedural action) (Oglesby et al. 2018). Coupled with skill acquisition and biomechanical enhancement, physiologic changes occur in major systems including the cardiovascular, respiratory, musculo-skeletal, and autonomic nervous systems (Sherwood 2010). Functional neural networks are also sculpted and modified by the training effects experienced by elite performers. These major networks include: (1) default mode network (responsible for internal focus such as thinking about the self and others and reflecting on the past and future) (Andrews-Hanna et al. 2014), (2) salience network (integrates internal and external stimuli and integrates sensory stimuli with visceral autonomic and emotional information) (Seeley et al. 2007), (3) central executive network (responsible for working memory, problem-solving, reasoning, task flexibility, and decision-making) (Seeley et al. 2007), (4) action-oriented network (coordinates action-execution and action-perception) (Krause et al. 2010), and (5) reward circuit network (formulates representations of something desired, predicts possibilities to acquire it and anticipates it) (Wittmann et al. 2007).

During repeated training, a sense of an embodied self is modified and refined. Accuracy of kinesthetic and proprioceptive feedback is internalized in order to perform at an elite level (Ericsson 2014). Long hours of mindful-focused practice are required to achieve physical expertise. When someone attains this level of expertise, teachers, coaches, and viewers immediately recognize the unique embodiment of a professional athlete or performing artist (Ericsson 2014). For individuals who reach this level of expertise, a highly refined neuromuscular feedback system operates to guide them towards peak performance, including an awareness of when they are performing at non-optimal levels. Along with training a highly refined interoceptive embodied self, expertise in embodied awareness must expand beyond the narrow boundary of the self. Embodied cognition enables rapid and accurate assessment regarding the intentions of others, both their physical movements as well as their social and emotional intentions. The capacity to experience empathy and theory of mind are deeply embedded within embodiment (Sevdalis and Keller 2011; Sevdalis and Raab 2014). Social contexts, temporal-spatial requirements to navigate effectively, and the biological movements and physiological states of others are all understood within our own functional neural networks. These complex networks decode internal and external information, which provides immediate implicit understanding about the actions and intentions of others (Grafton 2009).

Through intense training enduring neurological changes result. This is evidenced in a myriad of studies examining differences in brain structures in musicians, singers, dancers, and athletes compared to control subjects (Krause et al. 2010; Lotze et al. 2003). Acquiring physical expertise also influences stronger awareness of emotional and autonomic responses in self and other (Jackson and Decety 2004). Consequently, training promotes deeper perceptual understanding of bodily sensations, visceral/somatic activations, and self-awareness (Sze et al. 2010). For example, the notion of muscle memory is actually a physiologic reality. Muscles that are trained retain nuclei within the muscle fibers. Even after years of inactivity, a trained muscle will quickly adapt to conditioning demands and return to pre-atrophy states (Schwartz 2019). This physiologic response suggests that a trained embodied awareness in an elite performer is retained at the muscular level.

The effects of training may also cause damage to physiologic and neurological systems. Many elite performers suffer dystonias; neuromuscular signals are disrupted and performance ability declines, and for some performers or athletes, these highly trained motor patterns are no longer restored after injury and detraining (Geyer and Bressman 2006). The effects of stress, whether from physical fatigue or psychological stress, also negatively influence an embodied self (O'Donnell et al. 2008). The neurological stress system causes changes to the immune, autonomic, and neuromuscular systems. The capacity to perform at an elite level is undermined by acute and chronic stress (Baumann and Turpin 2010). As well, injuries change motor patterns; these changes also influence an embodied sense of self (Guptill 2012). Many elite performers will cultivate mental imagery and mental practice in order to sustain motor efficiency, especially when they are injured (Keller 2012). When optimal performance is diminished due to physical or neurological injuries or disorders, or by the negative influence of psychological disorders such as anxiety, depression, or posttraumatic stress disorder, rehabilitation programs are essential, including traditional motor re-patterning therapies. As well, rehabilitation programs continually incorporate new assistive technology treatment protocols (Baert et al. 2012; Leijnse et al. 2015).

8.4 Embodiment in Creativity

Embodiment is also evident in creativity. It is an integral component operating in the person or people who engage in the construction of a creative product. Embodied cognition is engaged in exploration, problem solving, idea generation, elaboration, critical thinking, and aesthetic orientation (Watts et al. 2017). Malinin (2019) proposes that understanding embodiment requires a radical investigative approach. Embodied cognition, and by extrapolation, embodied creativity, must include four dynamic components: (1) embodied (implicit body awareness), (2) embedded (coupling of physical and sociocultural environments), (3) enactive (meaningful actions), and (4) extended cognition (thinking is distributed beyond the body-mind of the individual) (Malinin 2019). Embodied creativity is a dynamic system, as evidenced in research studies that investigate distributed creativity. It expands beyond the individual person. When creativity is distributed it involves the creative contributions of many and these interactions operate at the implicit and explicit levels (i. e., group improvisation). Fundamentally, an embodied self is rooted in perceptions and phenomenological experiences that influence cognitive and creative appraisals; they operate simultaneously within the self, and expand to others, and the material world (Adams 2010).

Distributed embodied creativity is present in sport and the performing arts. Study findings demonstrate that physical exercise, such as aerobic exercise, has a direct positive effect on the creative process and creative output (Blanchette et al. 2005). Settings where motor activities include opportunities to freely explore movement ideas also promote creative expression and creative productivity (Bowers et al. 2014; Richard and Lebeau 2018). Mobilizing the body literally mobilizes the exploration of ideas and the persistence to engage in these explorations. Even acute bouts of aerobic activity, regardless of fitness level and training experiences, have positive effects on creativity (Roman et al. 2018). Perhaps the positive effects of exercise might also contribute to creativity within elite athletes and performing artists; their ability to access heightened creative output may actually operate as a feedback loop that supports physical and expressive abilities.

8.5 Creativity in Education and Technology

In educational settings, the classic systems models of creativity (i.e., distributed, socio-cultural) are frequently incorporated into educational programs and policies (Silvia et al. 2009). As well, the students and teachers bring their own embodied histories to the learning process and all of these factors converge and interface with educational technology (Shubina and Kulakli 2019). Generally, when students and teachers are exposed to technology such as digital platforms, where they acquire skills and confidence to navigate these platforms, increased exploration and creative processing manifest (Henriksen et al. 2016). When these tools are introduced effectively, intrinsic motivation and self-efficacy may increase in the students as well as the educators. Frequent positive exposure to educational technology has the potential to enhance problem-solving skills. As these skills are internalized, a sense of self-efficacy develops (Shubina and Kulakli 2019). Effective use of technology can promote novelty, usefulness, and an integration of context or wholeness into the learning experiences of students (Henriksen et al. 2016). The creative process can emerge within the individual or within a group of students, and this can take place in a diverse range of domains (science, arts, law, sport). When technology is supportive it can also heighten expertise as well as offer opportunities to interact with leaders in the field, a process that expands far beyond the walls of home, schools, and communities.

Given research on creativity and embodiment, the interface of interactive technology coupled with physical activity may stimulate more creative output (Levisohn and Schiphorst 2011). Embodiment is cultivated by interactions with the physical environment, and this includes interactions with digital environments (Farr et al. 2012). Technological advancements have facilitated new platforms to interface with embodied processing and cognition (Price et al. 2009). Some of the major technological modes of interaction include ubiquitous computing (information processing embedded into everyday artifacts and environments), proximity interfaces (uses the human body as a medium to store, display, and exchange information, i.e., mobile phones), wearable interfaces (worn on the human body and often considered extensions of the human body, i.e., headsets), sensory-specific interfaces (relies on human senses to activate the devices), implantable interfaces (medical devices, i.e., pacemakers, deep brain stimulators), and multi-modal interfaces (i.e., communal engagement, live streaming) (Price et al. 2009). Fundamentally, embodied cognition and creativity are driven by physical interactions, whether that is in real physical spaces or virtual spaces. Strong evidence suggests that technology is now deeply embedded within the structuring and functioning of cognitive/creative behavior. Multi-touch sensor and mobile technologies provide bodily-based physical experiences in novel delivery approaches (Farr et al. 2012). Interactive technology is increasingly integrated into all levels of educational programing, assessment, and infrastructure (U. S. Department of Education 2017).

8.6 Technology and Elite Performance

Exploring new technological tools is no longer a novel activity in the lives of students and educators. Technology is also firmly established in elite sport and performance programs (U. S. Department of Education 2017). Generally, technology intersects within three major aspects of sport and performance: (1) training (i.e., data collected to enhance physiologic and biomechanical fitness and increase skill level), (2) performance (i.e., improving clothing, footwear, equipment and performance surfaces/settings), and (3) judging, officiating and producing of an event (Fuss et al. 2008). Critical factors associated with technology must be incorporated when working with elite performers. Because technological devices are highly sophisticated, many professional organizations employ directors of technology to help coaches/athletes and producers/performers (Fuss et al. 2008).

The need for technological experts is evident in training practices such as when augmented feedback is incorporated. Augmented feedback includes a diverse range of devices and has enhanced training, motor learning strategies, and performance (Giblin et al. 2016). The advantage of sophisticated technology is vast; however, one essential limitation remains, and that is the ability of the athlete or performer to absorb the information and implement new motor behaviors. Technology must not overwhelm the participant; these assistive tools rapidly generate vast amounts of data and it is challenging for athletes and performers, as well as their coaches and teachers, to assimilate all this data into physical practice (Giblin et al. 2016). The timing of feedback is critical, especially in elite populations that have acquired highly refined skills. Educators and coaches must coordinate with the technology director and together they must address feedback practices that will account for the readiness of the athlete or performer (Gulhane 2014). The information must be both relevant to task-specific activities and meaningful to the athlete or performer.

One type of technology that has seen rapid adoption in elite settings, as well as in the general population, is wearable technology, such as Fitbits and smart watches. These devices can gather and relay real-time information about body functioning and body positions, as well as collect or send information via the Internet (Sandall 2016). The wearable sensors can be positioned on joints and limbs, and can guide and enhance motor behavior and optimize fitness training. They can also enhance cognitive processing such as attention and focus. Increasing effective cognitive processes can aid elite athletes and performers; this technology can also support individuals with learning disabilities regardless of whether they are elite performers or members of the general public. As well, wearable technology is fully integrated within the medical field. For example, medical devices can be surgically placed within the brain and/or body in individuals with cardiovascular or motor disorders. These surgically embedded sensors can monitor movement and regulate motor and physiologic functions (i.e., deep brain stimulation, cardiac pacemaker).

Beyond the physical responses that are enhanced by technology, accessing the virtual realm opens up new options. Assistive technology has operationalized the helpful practice of mental training by introducing mental imagery into a virtual space. It has expanded mental imagery practices and moved this passive form of mental training much closer to the ideal process of learning by doing (Nyberg et al. 2006). Wearable technology facilitates access to these virtual interactive realities. These innovations also expand options for individuals with physical disabilities; exploring these unique experiences is no longer limited to able-bodied people. Virtual realities delivered by wearable technology are increasing opportunities to participate, both as a performer and a spectator, such as in virtual games or virtual performances/sporting events. This form of technology has enhanced spectator/audience enjoyment and created safer environments for all who may participate in virtual reality games or performances.

All of these positive advances are equally vulnerable to malevolent use and so ethics committees and associations are now required to oversee the implementation of technological advancements in sport and performance settings (Omoregie 2016). There are significant concerns associated with technology and the wellness of elite athletes and performers. Many experts in the field fear a loss of self-awareness and self-discipline. For example, young children are now observing and then exploring advanced physical skills long before they have received sufficient training to execute these skills correctly. In a feature article in Dance Magazine (July, 2018), Howard captured the concerns of dance educators as well as accomplished professional dancers. The attention span and practice time of young dancers has diminished and been replaced by regular checking on social media platforms. The extreme images of thin hyper-mobile dancers (images that are often not accurate portrayals of the dancer or the movement) are propelling young or inexperienced dancers to emulate this aesthetic to the detriment of their physical and psychological health. The challenge for educators is to implement effective policies that will promote optimal utilization of technology, including modeling the integration of technology into classroom settings and programs (Henriksen et al. 2016). The educators, coaches, athletes, and performers must make wise choices related to technology and how it will be used to enhance performance, and most importantly, the embodied creative self (Omoregie 2016).

Beyond the positive and negative effects of technology, a profound shift has occurred in the traditional gatekeeping of expert assessment and guidance. With the increase in digital media and Internet content, the globalization and diversification of elite performance has re-located the audience. It is both the participant and the major determiner of media influence (Henriksen et al. 2016). Content sharing and development is rapidly transforming how culture is explored and perceived. Generally, this access to seemingly unlimited resources has accelerated exposure to new ideas and has supported rapid acquisition of information that can be directed towards novel and useful enterprises or products. It has also expanded the malevolent use of creativity; promotion of factitious ideas and racism has challenged expert and novice consumers who create digital media. Uninformed, poorly educated, or maliciously intended consumers are equal participants in this expanded virtual world (de Saint Laurent et al. 2020). Despite these negative occurrences, digital media has also introduced a plethora of new talent and expanded novel idea generation beyond the confines of pre-technology distribution.

8.7 Elite Athletes and Performers During a Pandemic

The COVID-19 pandemic witnessed the closure of all sporting and performing arts events. Simultaneously, technological platforms provided greater access to training and community digital events despite social isolation. Efforts to bring global communities together were facilitated by technological platforms; however, at the individual level, isolation produced both boredom and fatigue (Vallee 2020). Social isolation profoundly disrupts autonomic regulation, which negatively influences an empowered embodied sense of self (Grippo et al. 2007). This was evident during the pandemic. Social isolation was a significant complication for individual mental and physical wellbeing. Many children, adolescents and adults suffered anticipatory grief; they feared losing a loved one or dying from the virus. With increased need for medical and funeral services, technological tools offered important bridges. Telemedicine and telepsychology became medically responsible and viable services. Virtual funerals emerged to help families connect; they were able to engage in remote bereavement rituals. Although these virtual services offset the risk of escalating the pandemic, many were left struggling with feelings of disenfranchisement, and to some degree, disembodiment, from the traditional sociocultural connections that offer meaning and comfort (Brooks et al. 2020).

During the pandemic, the coronavirus took center stage and everything that was once normal was recalibrated. In the performing arts, 95% of the arts were canceled, 65% of the arts organizations reduced staff and artistic personnel, and yet 65% of these organizations still attempted to deliver content to the community via virtual technology (Wroblewski 2020). New ways to sustain a presence included exploring social platforms and integrating platforms such as Zoom into daily life. For example, performing arts training programs continued to offer classes virtually, and technological tools were rapidly integrated into arts education. Despite these sustained educational opportunities, after the initial novelty of this form of training, students struggled to stay engaged, although they appreciated the ease of distance learning (Davis 2020; Lempinen 2020). After the initial changes during the pandemic, several challenges emerged including "zoom fatigue", loss of in-person interactions and non-verbal communication, and diminished physical and social activities (Wroblewski 2020). Efforts to return to performance and studio venues were fraught with challenges and increased resurgence of the spread of the virus. This uneven and uncertain return intensified despair, and for some, denial propelled dangerous behaviors that increased risk. Seeking self-expression and reaching out to others to break isolation were efforts to offset the devastation of temporary or permanent loss of competitions, performance engagements, and even careers.

A slightly different pattern emerged in sport. A major shift from mainstream sports to e-sports moved sports' fans to a video-game version of their favorite sports. Some sports were more readily able to transition to simulations, for example formula one car racing, whereas other sports suffered without the immediacy of physical real-time play (Heaven 2020). Many sport fans were able to participate in virtual sports such as competitions between amateur cyclists who "entered" the mountainous virtual Tour de France race. These virtual competitions expanded opportunities to appreciate sport during the pandemic (Ovide 2020). Despite these technological innovations, professional and Olympic sport organizations suffered extensive financial losses and athletes lost valuable time to maintain fitness levels essential for peak performance. Efforts to keep elite athletes healthy and safe required collaboration with a wide range of participants and organizations including athletes, managers, agents, coaches, owners, and sporting associations (Ovide 2020). Similar to the performing arts sector, massive personal and financial losses were experienced and an uneven and uncertain return to sporting events persisted throughout the pandemic (Gough 2020). During this time, technological advancements provoked a reimagined fan experience and offered opportunities to globally support the health and wellbeing of professional and amateur athletes (Gough 2020).

The lived-isolation experienced during the pandemic revealed new research needs. According to Vallee (2020), more research is needed to understand how data is absorbed into the body during physical isolation. Fears, frustrations, anger, boredom, confusion, and stress (financial, psychological, anticipatory grief) all influence an embodied sense of self. Performances and training programs for elite performers, athletes and the general public were offered virtually. Virtual spaces were provided to connect, observe, and participate. Physical bodies participated in these virtual locations. This was a unique phenomenon that engaged embodiment and creativity. These activities helped reduce a sense of self in isolation (Price et al. 2009). Physical and virtual spaces converged; the intersection of these two spaces rapidly became a new normal. Reliance on technology, although helpful and meaningful, was, in reality, a poor substitute for face-to-face connections during this pandemic. The protracted quarantine process highlighted the limitations of technology; performers and athletes desired real-time connections and international engagements (Brooks et al. 2020).

8.8 Conclusion

Embodiment, creativity, and technology will continue to interact. Although we are constrained by the kind of body we possess (Borghi and Cimatti 2010), technology is opening pathways to expand the perceived limits of our physical capacities. With the expansion of our perceptions, an expanded embodied self can emerge. Just as we are free from our physical constraints in our dream life, technology is now supporting our imaginative yearnings to fly or enter alternate realities. These are exhilarating possibilities and yet we must also engage in real-time physical spatial domains. Highly trained athletes and performers have experienced the infectious exhilaration of flow and moments of awe. They are embodied states that reside within our interoceptive and exteroceptive sensorimotor systems (Csikszentmihalyi 1990).

As research continues to explore embodiment, growing awareness regarding the fluid boundaries of an embodied self must be operationalized. The fundamental concept of embodiment remains rich and valuable but theories must evolve to grasp how and when embodiment directly shapes behavior and cognition (Willems and Francken 2012). Embodied cognition and creativity far exceed the bounds of the human body. A large kinesphere enables us to capture microscopic and macroscopic signals from the environment. This can expand beyond the temporal and spatial bounds of the present (Wilson and Golonka 2013). Paranormal experiences can be included into embodied research (Yurtsever and Tasa 2009). As technological tools emerge, we can understand some of the more profound awe inspiring, mystical, and "superhuman" dimensions that currently remain mysterious or untestable (Holloway

2003; Stepanova et al. 2019). These exciting possibilities will be realized as immersive technological resources develop. However, the utilization of movement that is both task-oriented and experience-dependent must be preserved. Sensorimotor processes are vital and they are optimally developed in educational and training settings. These embodied experiences allow us to thrive in our sociocultural world. As well, an embodied self is readily able to interact with technology. Collectively, education, technology, and creativity offer the potential for physical and psychological wellbeing at the individual and societal level (Levisohn and Schiphorst 2011).

References

Adams, F. (2010). Embodied cognition. *Phenomenology of Cognitive Science, 9*, 619–628. https://doi.org/10.1007/s11097-010-9175-x

Andrews-Hanna, J. R., Smallwood, J., & Spreng, R. N. (2014). The default network and self-generated thought: Component processes, dynamic control, and clinical relevance. *Annals of the New York Academy of Sciences, 1316*(1), 29–52. https://doi.org/10.1111/nyas.12360

Baert, S., Casier, A., & De Raedt, R. (2012). The effects of attentional training on physiological stress recovery after induced social threat. *Anxiety, Stress and Coping, 25*(4), 365–379. https://doi.org/10.1080/10615806.2011.605122

Baumann, N., & Turpin, J-C. (2010). Neurochemistry of stress: An overview. *Neurochemistry Research, 35*, 1875–1879. https://doi.org/10.1007/s11064-010-0298-9

Blanchette, D. M., Ramocki, S. P., O'del, J. N., & Casey, M. S. (2005). Aerobic exercise and creative potential: Immediate and residual effects. *Creativity Research Journal, 17*(2 & 3), 257–264. https://doi.org/10.1207/s15326934crj1702&3_10

Borghi, A., & Cimatti, F. (2010). Embodied cognition and beyond: Acting and sensing the body. *Neuropsychologia, 48*, 763–773. https://doi.org/10.1016/j.neuropsychologia.2009.10.029

Bowers, M. T., Green, B. C., Hemme, F., & Chalip, L. (2014). Assessing the relationship between youth sport participation settings and creativity in adulthood. *Creativity Research Journal, 26*(3), 314–327. https://doi.org/10.1080/10400419.2014.929420

Brooks, S. K., Webster, R. K., Smith, L. E., Woodland, L., Wessely, S., Greenberg, N., & Rubin, G. J. (2020). The psychological impact of quarantine and how to reduce it: A review of the evidence. *The Lancet, 395*, 912–920. https://doi.org/10.1016/S0140-6736(20)30460-8

Corazza, G. E., & Glăveanu, V. P. (2020). Potential in creativity: Individual, social, material perspectives, and a dynamic integrative framework. *Creativity Research Journal, 32*(1), 81–91. https://doi.org/10.1080/10400419.2020.1712161

Critchley, H. D. (2009). Psychophysiology of neural, cognitive and affective integration: fMRI and autonomic indicants. *International Journal of Psychophysiology, 73*, 88–94. https://doi.org/10.1016/j.ijpsycho.2009.01.012

Csikszentmihalyi, M. (1990). *Flow: The psychology of optimal experience*. Harper & Row.

Davis, T. C. (March 23, 2020). Teaching performing arts during the pandemic. https://howlround.com/teaching-performing-arts-during-pandemic

de Saint Laurent, C., Glăveanu, V., & Chaudet, C. (2020). Malevolent creativity and social media: Crating anti-immigration communities on Twitter. *Creativity Research Journal, 32*(1), 66–80. https://doi.org/10.1080/10400419.2020.1712164

Dietrich, A. (2007). Who's afraid of a cognitive neuroscience of creativity? *Methods, 42*, 22–27. https://doi.org/10.1016/j.ymeth.2006.12.009

Dietrich, A. (2008). Imaging the imagination: The trouble with motor imagery. *Methods, 45*, 319–324. https://doi.org/10.1016/j.ymeth.2008.04.004

Ericsson, K. A. (2013). Training history, deliberate practice and elite sports performance: An analysis in response to Tucker an Collins review – what makes champions? *British Journal of Sports Medicine, 47*, 533–535.

Ericsson, K. A. (2014). Why expert performance is special and cannot be extrapolated from studies of performance in the general population: A response to criticism. *Intelligence, 45*, 81–103. https://doi.org/10.1016/j.intell.2013.12.001

Ericsson, K. A., Roring, R. W., & Nandagopal, K. (2007). Giftedness and evidence for reproducibly superior performance: An account based on the expert performance framework. *High Ability Studies, 18*, 3–56. https://doi.org/10.1080/13598130701350593

Farr, W., Price, S., & Jewitt, C. (2012). An introduction to embodiment and digital technology research: Interdisciplinary themes and perspectives. *National Centre for Research Methods Working Paper, Institute of Education, London.* http://eprints.ncrm.ac.uk/2257/

Fuchs, T., & Koch, S. C. (2014). Embodied affectivity: On moving and being moved. *Frontiers in Psychology, 5*, 508. https://doi.org/10.3389/fpsyg.2014.00508

Fuss, F. K., Subic, A., & Mehta. (2008). The impact of technology on sport: New frontiers. *Sports Technology, 1*(1), 1–2, https://doi.org/10.1080/19346182.2008.9648443

Gaggioli, A., Mazzoni, E., Benvenuti, M., Galimberti, C., Bova, A., Brivio, E., Cipresso, P., Riva, G., & Chirico, A. (2020). Networked flow in creative collaboration: A mixed method study. *Creativity Research Journal, 32*(1), 41–54. https://doi.org/10.1080/10400419.2020.1712160

Geyer, H. L., & Bressman, S. B. (2006). The diagnosis of dystonia. *Lancet Neurology, 5*, 780–790. https://doi.org/10.1016/S1474-4422(06)70547-6

Gough, C. (Jun 26, 2020). Coronavirus (COVID-19) disease pandemic effect on the sports industry – Statistics and facts. https://www.statista.com/topics/6098/impact-of-the-coronavirus-on-sport/

Giblin, G., Tor, E., & Parrington, L. (2016). The impact of technology on elite sports performance. *Sensoria – A Journal of Mind, Brain and Culture, 12*(2), 3–9. https://doi.org/10.7790/sa.v12i2.436

Grafton, S. T. (2009). Embodied cognition and simulation of action to understand others. *Annuals of New York Academy of Science, 1156*, 97–117. https://doi.org/10.1111/j.1749-6632.2009.04425.x

Grippo, A. J., Lamb, D. G., Carter, C. S., & Porges, S. W. (2007). Social isolation disrupts autonomic regulation of the heart and influences negative affective behaviors. *Biological Psychiatry, 62*, 1162–1170. https://doi.org/10.1016/J.biopsych.2007.04.011

Gulhane, T. F. (2014). Various types of advanced technologies in sports. *Journal of Sports and Physical Education, 1*(6), e-ISSN: 2347–6737

Guptill, C., (2012). Injured professional musicians and the complex relationship between occupation and health. *Journal of Occupational Science, 19*(3), 258–270. https://doi.org/10.1080/14427591.2012.670901

Heaven, W. D. (June 22, 2020). The pandemic will change how we watch sports. https://www.technologyreview.com/2020/06/22/1004194/lockdown-pandemic-sports-esports-virtual-fans-watch-video-games/

Henriksen, D., Punya, M., & Risser, P. (2016). Infusing creativity and technology in 21st Century education: A systemic view for change. *Educational Technology and Society, 19*(3), 27–37. https://creativecommons.org/licenses/by-nc-nd/3.0/

Herbert, B. M., & Pollatos, O. (2012). The body in the mind: On the relationship between interoception and embodiment. *Topics in Cognitive Science, 4*, 692–704. https://doi.org/10.1111/j.1756-8765.2012.01189.x

Holloway, J. (2003). Make-Believe: Spiritual practice, embodiment, and sacred space. *Environment and Planning A: Economy and Space, 35*(11), 1961–1974. https://doi.org/10.1068/a3586

Howard, T. R. (2018, July 16). Is Instagram changing the dance world's value system? *Dance Magazine*. https://www.dancemagazine.com/social-media-and-dance-2585216791.html?rebelltitem=7#rebelltitem7

Jackson, P.L., & Decety, J. (2004). Motor cognition: A new paradigm to study self-other interactions. *Current Opinion in Neurobiology, 14*, 259–263. https://doi.org/10.1016/j.conb.2004.01.020

Keller, P. E. (2012). Mental imagery in music performance: Underlying mechanisms and potential benefits. *Annals of New York Academy of Sciences, 1252*, 206–213. https://doi.org/10.1111/j.1749-6632.2011.06439.x

Koziol, L. F., Budding, D. E., & Chidekel, D. (2012). From movement to thought: Executive function, embodied cognition, and the cerebellum. *Cerebellum, 11*, 505–525. https://doi.org/10.1007/s12311-011-0321-y

Krause, V., Schnizler, A., & Pollok, B. (2010). Functional network interactions during sensorimotor synchronization in musicians and non-musicians. *Neuroimage, 52*, 245–251. https://doi.org/10.1016/j.neuroimage.2010.03.081

Lakoff, G. (2012). Explaining embodied cognition results. *Topics in Cognitive Science, 4*, 773–785. https://doi.org/10.1111/j.1756-8765.2012.01222.x

Leijnse, J. N. A. L., Hallett, M., & Sonneveld, G. J. (2015). A multifactorial conceptual model of peripheral neuromusculoskeletal predisposing factors in task-specific focal hand dystonia in musicians: Etiologic and therapeutic implications. *Biological Cybernetics, 109*, 109–123. https://doi.org/10.1007/s00422-014-0631-5

Lempinen, E. (May 27, 2020). The pandemic could open a door to new technology – and dramatic innovation – in education. https://news.berkeley.edu/2020/05/27/the-pandemic-could-open-a-door-to-new-technology-and-dramatic-innovation-in-education/

Levisohn, A., & Schiphorst, T. (2011). Embodied engagement: Supporting movement awareness in ubiquitous computing systems. *Ubiquitous Learning, 3*(4), 97–111. http://ijq.cgpublisher.com/, ISSN 1835-9795

Llinas, R. R. (2001). *I of the vortex: From neurons to self.* Cambridge, MA: The MIT Press.

Lotze, M., Scheler, G., Tan, H-R. M., Braun, C., & Birbaumer, N. (2003). The musician's brain: Functional imaging of amateurs and professionals during performance and imagery. *Neuroimage, 20*, 1817–1829. https://doi.org/10.1016/j.neuroimage.2003.07.018

Malinin, L. H. (2019). How radical is embodied creativity? Implications of 4E approaches for creativity research and teaching. *Frontiers in Psychology, 10*, 2372. https://doi.org/10.3389/fpsyg.2019.02372

Montero, B. (2006). Proprioception as an aesthetic sense. *Journal of Aesthetics and Art Criticism, 64*(2), 231–242. https://doi.org/10.1111/j.0021-8529.2006.00244.x

Nyberg, L., Eriksson, J., Larsson, A., & Marklund, P. (2006). Learning by doing versus learning by thinking: An fMRI study of motor and mental training. *Neuropsychologia, 44*, 711–717. https://doi.org/10.1016/j.neuropsychologia.2005.08.006

O'Donnell, K., Brydon, L., Wright, C. E., & Steptoe, A. (2008). Self-esteem levels and cardiovascular and inflammatory responses to acute stress. *Brain, Behavior, and Immunity, 22*, 1241–1247. https://doi.org/10.1016/j.bbi.2008.06.012

Oglesby, C., Henige, K., McLaughlin, D., & Stillwell, B. (Eds). (2018). *Foundations of Kinesiology.* Jones & Bartlett Learning.

Olszewski-Kubilius, P., Subotnik, R. F., & Worrell, F. (2015). Antecedent and concurrent psychosocial skills that support levels of achievement within talent domains. *High Ability Studies, 26*(2), 195–210. https://doi.org/10.1080/13598139.2015.1095077

Omoregie, P. O. (2016). The impact of technology on sport performance. *Proceedings of INCEDI 2016 Conference,* 896–905. ISBN: 978-9988-2-3994-7

Ovide, S. (July 10, 2020). Sports in a pandemic don't all stink. https://www.nytimes.com/2020/07/10/technology/virtual-cycling.html

Perlovsky, L. (2015). Origin of music and embodied cognition. *Frontiers in Psychology, 6*(538), 1–4. https://doi.org/10.3389/fpsyg.2015.00538

Price, S., Roussos, G., Falcao, T. P., & Sheridan J. G. (2009). Technology and embodiment: Relationships and implication for knowledge, creativity and communication. *Beyond Current Horizons: Children, schools and families*. https://www.beyondcurrenthorizons.org.uk

Richard, V., & Lebeau, J-C. (2018). Developing cognitive and motor creativity in children through an exercise program using nonlinear principles. *Creativity Research Journal, 30*(4), 391–401. https://doi.org/10.1080/10400419.2018.1530913

Rietveld, E. (2008). Situated normativity: The normative aspect of embodied cognition in unreflective action. *Mind, 117*, 973–1001. https://doi.org/10.1093/mind/fzn050

Roman, P. A. L., Vallejo, A. P., & Aguayo, B. B. (2018). Acute aerobic exercise enhances students' creativity. *Creativity Research Journal, 30*(3), 310–315. https://doi.org/10.1080/10400419.2018.1488198

Sandall, B. K. (2016). Wearable technology and schools: Where are we and where do we go from here? *Journal of Curriculum, Teaching, Learning, and Leadership in Education, 1*(1), Article 9. https://digitalcommons.unomaha.edu/ctlle/vol1/iss1/9

Schwartz, L. M. (2019). Skeletal muscles do not undergo apoptosis during either atrophy or programmed cell death: Revisiting the myonuclear domain hypothesis. *Frontiers in Physiology 9*, 1887. https://doi.org/10.3389/fphys.2018.01887

Seeley, W. W., Menon, V., Schatzberg, A. F., Keller, J., Glover, G. H., Kenna, H., Reiss, A. L., & Greicius, M. D. (2007). Dissociable intrinsic connectivity networks for salience processing and executive control. *Journal of Neuroscience 27*, 2349–2356. https://doi.org/10.1523/JNEUROSCI.5587-06.2007

Sevdalis, V., & Keller, P. E. (2011). Captured by motion: Dance, action understanding, and social cognition. *Brain and Cognition, 77*, 231–236. https://doi.org/10.1016/j.bandc.2011.08.005

Sevdalis, V., & Raab, M. (2014). Empathy in sports, exercise, and the performing arts. *Psychology of Sport and Exercise, 15*, 173–179. https://doi.org/10.1016/j.psychsport.2013.10.013

Sherwood, L. (2010). *Human physiology from cells to systems* (7th ed.). Brooks/Cole.

Shubina, I., & Kulakli, A. (2019). Pervasive learning and technology usage for creativity development in education. *International Journal of Emerging Technologies in Learning, 14*(1), 95–109. https://doi.org/10.3991/ijet.v14i01.9067

Silvia, P. J., Kaufman, J. C., & Pretz, J. E. (2009). Is creativity domain-specific? Latent class models of creative accomplishments and creative self-descriptions. *Psychology of Aesthetics, Creativity, and the Arts, 3*(3), 139–148. https://doi.org/10.1037/a0014940

Stepanova, E. R., Quesnel, D., & Riecke, B. E. (2019). Understanding AWE: Can a virtual journey, inspired by the overview effect, lead to an increased sense of interconnectedness? *Frontiers in Digital Humanities, 6*, 9. https://doi.org/10.3389/fdigh.2019.00009

Sze, J. A., Gyurak, A., Yuan, J. W., & Levenson, R. W. (2010). Coherence between emotional experience and physiology: Does body awareness training have an impact? *Emotion, 10*(6), 803–814. https://doi.org/10.1037/a0020146

U. S. Department of Education (2017). *Reimaging the role of technology in education: 2017 National education technology plan update*. http://tech.ed.gov

Vallee, M. (2020). Doing nothing does something: Embodiment and date in the COVID 19 pandemic. *Big Data and Society*, January – June, 1–12. https://doi.org/10.1177/2053951720933930

Watts, L. L., Steele, L. M., & Song, H. (2017). Re-examining the relationship between need for cognition and creativity: Predicting creative problem solving across multiple domains. *Creativity Research Journal, 29*(1), 21–28. https://doi.org/10.1080/10400419.2017.1263505

Willems, R. M., & Francken, J. C. (2012). Embodied cognition: Taking the next step. *Frontiers in Psychology, 3*, 582. https://doi.org/10.3389/fpsyg.2012.00582

Wilson, A. D., & Golonka, S. (2013). Embodied cognition is not what you think it is. *Frontiers in Psychology, 4*, Article 58. https://doi.org/10.3389/fpsyg.2013.00058

Wittmann, B. C., Bunzek, N., Dolan, R. J., & Duzel, E. (2007). Anticipation of novelty recruits reward system and hippocampus while promoting recollection. *Neuroimage, 38*, 194–202. https://doi.org/10.1016/j.neuroimage.2007.06.038

Wroblewski, A. (June 2, 2020). The present and future of arts organizations, technology and the pandemic. https://amt-lab.org/blog/2020/5/the-present-amp-future-of-arts-organizations-technology-and-the-pandemic

Yurtsever, A. Tasa, U. B. (2009). Redefining the body in cyberculture: Art's contribution to a new understanding of embodiment. In D. Riha & A. Maj (Eds.), The real and the virtual: Critical issues in cybercultures (pp. 1–15). The Interdisciplinary Press.

Paula Thomson, PsyD, is Professor, California State University, Northridge and licensed Clinical Psychologist. She is Co-Director of the Performance Psychophysiology Laboratory and Professor Emeritus/Senior Scholar at York University (Canada). She conducts research investigating attachment, early trauma, and creativity. She is the co-author of two books, Creativity and the Performing Artist: Behind the Mask and Creativity, Trauma, and Resilience and author of multiple chapters and peer-reviewed articles. She was a professional dancer and continues work as choreographer and movement coach in dance, theatre and opera. In 2013, she was named one of the top 20 female professors in California.

S. Victoria Jaque, PhD, is a Professor at California State University, Northridge where she is the Director of two graduate programs. She is Co-Director of the Performance Psychophysiology Laboratory and conducts research addressing the intersection between psychopathology and the optimal activation of the autonomic nervous system during performance. She is the co-author of two books with Dr. Thomson, Creativity and the Performing Artist: Behind the Mask and Creativity, Trauma, and Resilience and co-author of multiple peer-reviewed articles. She also collaborates on work examining the efficacy of resistance training protocols for the optimization of bone adaptation during maturation.

Chapter 9
What Else Can This Be?: Creativity as an Iterative Practice

Rabbi Adina Allen and Pat B. Allen

Abstract This chapter describes a form of learning in which the technologies of creative engagement – art making, writing, singing and embodied focus – and a generative form of intellectual inquiry through text study that is centuries-old, are combined. The creative practice originates in the field of art therapy and the practice of engagement with text is a foundational feature of Judaism. Together these strands weave the Jewish Studio Process (JSP) – a psycho-spiritual technology of self and life inquiry that places the focus and direction of learning and creating in the hands of each participant. Though arising from specific disciplines – art therapy and Jewish spirituality – this chapter seeks to provide those of all art and religious backgrounds, and none, a way into this work. This chapter explores the various elements of this methodology as well as the digital technology adaptations that were made in the wake of the COVID-19 pandemic. We close with an invitation and support for how to bring these practices to different educational settings.

Keywords Art · Creativity · Spirituality · Education · Art therapy · Studio · Open studio · Judaism

9.1 Introduction

The work described in this chapter offers a form of learning in which the technologies of creative engagement – art making, writing, singing and embodied focus – and a generative form of intellectual inquiry through text study that is centuries-old are combined. The creative practice originates in the field of art therapy and the practice of engagement with text is a foundational feature of Judaism.[1]

[1] *Encyclopaedia Judaica*, 2008 The Gale Group.

R. A. Allen · P. B. Allen (✉)
Jewish Studio Project, Berkeley, CA, USA

D. Henriksen, P. Mishra (eds.), *Creative Provocations: Speculations on the Future of Creativity, Technology & Learning*, Creativity Theory and Action in Education 7,
https://doi.org/10.1007/978-3-031-14549-0_9

Together these strands weave the Jewish Studio Process (JSP) – a psycho-spiritual technology of self and life inquiry that places the focus and direction of learning and creating in the hands of each participant. This view of education is one that is learner-centered, democratized and holistic, such that all parts of the learner are welcome and necessary to the learning process. While this practice is situated in a specific spiritual context, we have found that all of the aspects are translatable to a wide variety of secular educational settings where creativity is the goal.

The foundation of the JSP methodology and philosophy is a belief in the Creative Source as something universal and available to everyone. Relying on this belief guided the mindset of opportunity that led to an expansion of services, a deepening of the methods used, and new insights into such topics as accessibility, autonomy and community building that can be translated into other learning situations, both online and in person.

The JSP is a method that enables full presence: mind, body, heart, soul and spirit to be activated in the service of deepening self-understanding, increasing communal resonance among those present, and opening new insights into societal issues. Through an understanding of the mind-body connection, the method traverses the expressive continuum to first build a shared resonance among those present and to engage each part of our sensing and apprehending possibilities, while placing the control of the experience gently in the hands of each person involved.

The structure of the method is a unique blend of individual and group education that is formal and predictable, allowing relaxation of the nervous system that is often activated in novel education and creative experiences. Yet within the formal structure, freedom of expression is available and invites something unforeseen and unpredictable to occur. Within every session, the method allows a period of activation balanced by segments of relaxation to increase resilience and to develop capacity to accept self-modulated challenges in response to group prompts. The goal is to offer maximum agency for those participating while adhering to the overall design. This creates a sense of safety, ease and belonging. At the same time, the value of curiosity and permission to reference what others are creating invites observation of fellow participants without disruptive comments. The combination of creating group resonance and encouraging individual curiosity results in a unique experience of safety (no comment) and porousness (notice everything) that effortlessly builds the skill of self regulation and organically encourages relationships to develop – to others, to one's self and to the material presented. When we see another take a risk on an art piece, say making bold marks in a strong color, or raise a wondering question rather than a right answer, those possibilities are suddenly open to us as well, creating more spaciousness and freedom. Yet, there is no pressure to create or respond in a way similar to any other person. Because the method helps each person center and observe the self while staying in relation to others, it becomes possible to "feel into" (Gendlin 1982) the experience I peripherally observe. Seeing someone make small, repetitive marks, attracts my energy, I may decide to try that myself.

9.2 Open Studio Process

The creative expression part of the experience (Open Studio Process) is an iteration of art therapy that grew out of the author's (Allen 1995, 2005) wrestling with certain accepted elements of the discipline that distinguish it as a clinical practice as it developed in the United States since the 1940s. The popular notion of marrying art making and psychotherapy arose alongside the discovery of the unconscious by Sigmund Freud. In Europe, psychiatrists considered the art produced by those suffering with mental illness fascinating but indecipherable (Prinzhorn 1966). Meanwhile, American psychologist Margaret Naumburg (1966), one of art therapy's founding pioneers, instead recognized such imagery as symbolic speech emanating from the inner life of each person. Naumburg's background in the free school movement of the 1930s led her to understand how self-expression creates self understanding and is the foundation of healthy autonomy (Naumburg 1928). Naumburg envisioned the role of the art therapist as an advocate for her patients and for the images they created, upholding the dignity and importance of their art as a means to understand and help alleviate their suffering.

Art therapy has grown into a profession closely tied to the medical model and grants a place in the hierarchy of care to the art therapist as interpreter of client art, emphasizing the psychotherapist aspect of the professional equation. While this approach serves the art therapist and the profession in the practical sphere of jobs and social placement, it can also prevent the client from fully claiming their agency as the primary interpreter of their own life and story.

The Open Studio Process (OSP) offers an alternative in which meaning is made by each person who participates, with the witness of others, including those who facilitate, who themselves are fellow travelers engaging in the same process. All are present as artists seeking to understand themselves and one another. The Open Studio Process is a formal practice that consists of three main steps undertaken by each individual:

- Setting of an **intention**, a statement that defines the parameters of the creative experience, both limits and aspirations
- Making of **art**, that is using materials to make marks or fashion objects
- Writing and reading of **witness** documents that describe and inquire into and with the image that emerges for each person.

The OSP also has a few simple rules: **Follow pleasure**, meaning to respect and embrace the energy created by creating. **Notice everything** invites ourselves to be in a relaxed awareness where, while our hands are engaged in making, our minds can notice what activates us, both the pleasant and the challenging. In this process we need not attend to that fully but rather allow those stimulating elements to be held in the creative process. **No-comment** asks all present to refrain from speaking any comments, questions, offering advice, interpretation or judgments, including any compliments, to one another during an OSP session. Inner comments and judgments about one's own work, which naturally arise in each person, are invited into the witness writing segment and can be read aloud or not, in whole or in part, during the designated time at the end of the session. Reading aloud is an option, but is never required. As we hear our own words, either internally if we read silently, or aloud, we become aware of the multivocal nature of our inner existence and how we have agency to choose what to focus on. We learn that marks of a certain color and duration serve to free us from difficult emotions while providing a satisfying sense of pleasure that rewards our attention.

During art making, those present are encouraged to engage in **referencing**, a techne of trying out or taking into one's own process a color, type of mark making or other technique they see others using. Most of us hold a memory of being told "not to copy" in school. And while peering over to copy answers on a test from another's page is objectionable, all art that has ever been created draws on **referencing**, from another close by, observation of nature and the world or from examples in art history. Encouraging the act of seeing and being inspired by the work of others helps to break down hierarchies and reinterprets creativity as a communal and collaborative process, even if the object created is individually wrought. Traces of reference in terms of marks and even, at times, themes, can be observed in groups. Rather than having a teacher do a demonstration that each student tries to replicate, each unique iteration of expression in the studio builds on and helps to create a community of images and ways of working that becomes an accessible shared world, a creative commons. We can all access all of the techniques and insights that are communally held and evolved. This ethos is echoed in another key feature of Jewish Studio Process taken from historical Jewish study, the *Beit Midrash*.

9.3 Beit Midrash

Beit Midrash is both a core practice and a place in the physical studio. *Beit* means house and *midrash* comes from the Hebrew word which means "to seek" or "inquire." The term *beit midrash* refers to both the traditional Jewish practice of inquiring into text, as well as the place in which this inquiry occurred. After the destruction of the 2nd Temple in Jerusalem in 70 CE, the system of sacrificial rites which constituted the core of Judaism could no longer be practiced and Judaism had to either be reinvented or die. Out of this rupture emerged the two pillars that have defined Judaism up until the present day – *beit kineset* (the house of prayer) and *beit midrash*. Liturgical prayer and rigorous engagement with sacred texts replaced the sacrificial system that had defined Judaism up until this point. Text study stood in place for the emotional, somatic, risky, intense experience of sacrifice – and was approached with that level of care, fierceness and import. In place of the complex nature of sacrifices arose an elaborate system for learning and commenting on sacred texts – and for preserving those commentaries. This allowed those engaged to carry on communal conversations across the generations and create a new canon to which Jews around the world continue to engage for challenge and guidance. A page of talmud is in itself a visually interesting document that serves to illustrate the principle of this conversation. While since about 70 CE *beit kineset* (synagogue prayer) has held center stage as the major location of Jewish practice, today *beit midrash* is re-emerging as a primary pathway for spiritual practice, connection and community.

Jewish Studio Project is at the forefront of the movement to reinvent beit midrash – what it means, the purposes it can serve, and whom it is for – in our day. While many who engage in *beit midrash* learning do so for the intellectual stimulation, to increase their Jewish literacy, or to learn practical laws and rules, JSP chooses to lift up other purposes this way of learning can serve. Our practice of beit midrash foregrounds this technology as a means of engaging relevant questions about self and the world through immersion in sacred ancient text; of building relationship – with the text, with one's learning partner, with the generations that came before, and with one's own self; and of building, preserving and reinventing community. Ultimately, for us, *beit midrash* is a practice through which each person who engages activates the prophetic voice within. Each of us has a piece of the wisdom that is needed to bring about a more just world. Beit Midrash can continue to preserve Jewish knowledge and community and, we imagine, has relevance to the wider human community in this time of fraught relationship and polarization. The key element to this wider application is the practice of *chevrutah.*

9.4 Chevrutah

We hold as central the ancient practice of *chevrutah* – engaging in the study of text with one other person or in a small group. Connected to the Hebrew word *chaver* (friend) and deriving from the Aramaic word *chibbur* (to bind together), *chevrutah* emphasizes the deeply relational nature of *beit midrash* and the act of textual inquiry as a dialogical, dialectical and multi-vocal process.

While for hundreds of years the doors to the *beit midrash* were only open to men, our version of this practice flings the doors wide open so that anyone coming with an open mind, curiosity and a willingness to engage is welcome. This includes all those who for so long have been left on the margins: women, those who are non-binary or gender non-confirming, Jews of color, queer folx, those who don't believe in God, those without formal Jewish education or literacy, and non-Jews. When we engage in text study in our programs we use the term *beit midrash* and *chevrutah* deliberately, choosing to expand the notion of what these words mean, including who is seen as a legitimate participant, what constitutes a text, which ideas are welcome, and the purpose of this learning itself. The results are profound shifts in the people who participate, an expansion of the tradition, and a deepening and expansion of what the texts can mean.

Elie Holzer and Orit Kent (2014) have studied the pedagogical implications and requirements of this way of approaching meaning making and the building of knowledge. Their research informs JSP's approach and has much to offer learning environments outside the Jewish world. As an educational technology, *beit midrash* is a means of intellectual engagement that makes space for – indeed requires – bringing one's own life experience, feelings, and beliefs into the learning. The practice of *chevrutah* in the way we employ it decentralizes the learning away from a model of the "expert" teacher at the front of the room who has the answers and asks, instead, for each pair or small group to parse out the text and come to their own insights. These insights are then brought into a group discussion following *chevrutah* where these ideas can come into contact with each other and learners begin to build a shared understanding of the world. The goal is to raise at least as many questions as answers by placing the text in conversation with poetry, contemporary writings as well as ancient commentaries. In this way *beit midrash* is a radical practice that democratizes learning and allows for new insights and truths to emerge freely, without being limited by the ideas or beliefs of the teacher.

We have found that the practice of Spiritual Grounding helps to prepare both the learners and the facilitator for the work of *beit midrash* and *chevrutah learning* to be most effective.

9.5 Spiritual Grounding

In any educational setting, a simple routine can serve the most basic requirements of forming a basis of predictability, comfort and warm up that helps prepare the learner to enter the learning environment and be present to the content being offered. We have found through practice that viewing this entry into the learning process as necessary not only for the learner, but for the educator/facilitator, as well as for the entire group together – creates the optimal environment.

Spiritual grounding is a way of coming into shared resonance and connecting to something bigger and beyond the self/deep within the self, depending on one's theological framework. We use spiritual grounding as a technology to begin the work of textual inquiry and art-making. Spiritual grounding can include blessing, singing, personal sharing, speaking to the realities of the present moment, movement, embodied meditation, contemplative silence, and more. We consider spiritual grounding a reimagined form of set induction necessary in education to provide readiness for new information. In our work, we're asking: What type of readiness is needed for the type of learning we are doing? Our answer is heart-opening, alignment of our being, a relaxation of the nervous system, an attunement to with whom we are in the room, both past and present, (i.e, the authors of the texts we are about to study) and an orientation towards the transformative purpose of this work.

9.6 Jewish Studio Process

The Jewish Studio Process (JSP) combines the three technologies of spiritual grounding, *beit midrash* and the Open Studio Process, into a new methodology. This Process invites the individual personal focus that adheres to OSP and the intellectual focus of beit midrash and engenders a cohering that situates the self along with others in the larger reality. We are living at a time in which we must do work both on the level of the individual and on the level of the community to create the alignment needed for cultural change to occur. In this time, collective consciousness is opening to realize our interconnection, and we are coming to see that our collective issues are only going to be solved through each of us doing this personal work. The Jewish Studio Process allows for both.

As an educational technology, this process not only recognizes the need for social and emotional learning in addition to the intellectual but also the importance of both of these learning processes being held together in the same space, done with the same group of people rather than bifurcating where and with whom one engages their learning, spirituality, social justice, and creativity. In this way, the JSP also brings together all parts of self: intellect along with intuition and imagination. As historian Karen Armstrong writes in her most recent book *The Lost Art of Scripture* (2020), we do violence when we only engage text through the left side of the brain. The time has come for the reintegration of these parts of self. Here, the left and right side of the brain are both given space at the table and room is made both for intellectual analysis, logic and rationality as well as for emotions, holistic thinking, bodily sensing, and empathy.

Additionally, core to the JSP as an educational technology is the concept of participant facilitation – that the values must be practiced by the practitioner. A JSP facilitator is at once a participant in *and* a facilitator of the Process. We call this a participant-facilitator. To facilitate literally means to make *facil* – easy or easeful. A JSP facilitator is one who helps others to use the Process to easefully access their own inherent wisdom. Rather than a "pedagogy" to be applied to one's teaching, this is something that must be lived and practiced. A facilitator of the Process creates the container for participants' – and their own – inner knowing to bubble up, and for a wisdom that is bigger and beyond the self to come down. The way a JSP facilitator creates this container is two-fold: (1) Through leading the steps and rules of the Jewish Studio Process, (2) By being authentically alive to his/her/their *own* process. Both of these aspects are essential. The first involves keeping time, giving clear instructions, paying attention to details of room set up and materials, articulating the values and purpose behind each part of the process. The second involves being alive and open to Source. One's connection to and belief in the Process is what allows others to come along on the journey.

9.7 Space

The space in which people learn is critical in educational settings. In classrooms teachers focus on bulletin boards, number charts, learning devices, inspiration and all the other factors that traditionally go into classroom set up. Yet, as we get older and move through the educational system, this aesthetic aspect often falls away to the point where college classes often take place in rooms in which the shape and color of the cinder blocks making up the walls are indistinguishable from a hospital or a prison, without even windows to the outside. The look, feel, and set up of space is a primary technology through which the JSP gets communicated.

The space that houses Jewish Studio Project is unique in combining both a freewheeling art space where individuals often paint standing at a wall, allowing traces of their work to remain visible after they finish and a library with shelves of books, including leather bound gold embossed volumes of Talmud and other sacred texts alongside contemporary books on Judaism, psychology, art history, feminism, and social justice. There is a couch and chairs in the library, floor pillows and carpeting that invite stretching out and reading in a relaxed state. The space flows into the studio section where tables fill the center space and shelves under the tall windows overflow with found objects, an array of drawing materials, and the basic tools to allow for many types of construction and expression. The design of the space is itself a technology that physically expresses the ethos of foundational JSP beliefs. When a participant places their paper on a section of wall near the books, they are already shifting their perception of what is possible. The space combines order and orderly chaos. Boxes of found objects are open for perusal and exploration, not labeled and closed. Colorful posters hand painted with the foundational texts hang on the walls above where participants engage in text study. The set up verifies the commitment to creative expression and subtly builds trust and engagement as soon as participants cross the threshold.

Until the COVID-19 pandemic, JSP was almost exclusively an embodied practice that took place in this art studio setting with groups of individuals in close proximity. The technologies involved were foundationally physical: sturdy paper or canvas hung on a wall, colorful paints, big brushes. While adjustments were always available to ensure accessibility for those unable to stand or work large, there has been a deep belief in the power of accepting and engaging the body as fully as possible (Gendlin 1982). Warm ups and grounding exercises include breath work, singing, moving meditations and guided imagination. The organizer is always a participant facilitator, both holding the space in terms of care of materials, time keeping and explaining the structure, and taking part in every step of the process. Being in a space and opening the self to one's own creative process in a community of practicing others allows for a kinesthetic education through immersion. As each person uniquely experiments with materials, others present learn by absorbing techne that they might not have used without having seen another person say, use a fan brush used to make feathery shapes, or what happens when someone goes over their painting with a big brush filled with water blending colors with gusto. All are present as artists.

The holding environment for the text discussions then, is an aesthetic, visually lively space. Snacks and drinks are always available. The continued presence of images made hanging on the walls allows the images themselves to be part of the process and the energy they emanate supports the back and forth between *chevruta* partners. These generative physical aspects of the Jewish Studio Process came to a halt, as so many other learning situations did, in March of 2020.

9.8 Digital Technology Adaptations

It is the commitment to JSP principles – trust the process, creativity is inherent in everyone, every challenge is an opportunity, the Divine is infused in everything – that allowed us not only to successfully adapt this method during the pandemic, but for JSP to grow, thrive and be of service during this time of challenge. As an organization centered around educational methodology facilitated at immersives, trainings, conferences, and gatherings, during the early weeks of the pandemic it was unclear how JSP would continue during a prolonged period of social distancing. The pandemic made us ask the questions: what else could this be? How else could this practice look? What is most needed now? Trusting in the creative process, we allowed ourselves to say goodbye (at least for a time) to the work as we had known it, to mourn what was no longer, and then to open to the creative process to find out what might come next. We began by offering virtual programs free of charge to our alumni and core community and then began to build out additional offerings from there. The first step was an act of care. We were amazed and delighted by the ways in which the JSP translated to the virtual format and to the impact possible during a time of global challenge, uprising and awakening. The deep appreciation expressed by those who attended programs helped guide and inspire us to do more.

The pandemic led us to adapt the JSP to a virtual setting using digital technology. We sought to create the look and feel of the in-person studio space via zoom, which we did in a number of ways. We were conscious of attending to transitions explicitly by designing sessions with a thoughtful energetic arc and using care when moving between one type of energy (e.g. heaviness, sadness, focus on global issues) to another (e.g. joy, relaxation, grounding). In order to enhance the feeling of connection, facilitators shared computer sound so that all in the program had the same music, creating a shared container. (The option to mute that sound and play one's own music is always permissioned). To create the visual feel of the studio, facilitators have their own artwork up on the walls behind them so that it can be seen during a session. This passive sharing of in-process and finished pieces helps invite creative possibilities for participants and conveys the value of participant facilitation. Additionally, during the art-making portion of the sessions, facilitators invite folks who would like to tilt their camera so that we can see each other's hands moving as they create. This both allows for a sense of feeling connected to other people even when you are alone in your own space, and opens the possibility for referencing, as mentioned earlier in this chapter. In terms of flow of the sessions, it has been important to focus on a multi-modal approach in order to retain energy and increase engagement in the virtual setting. Changing up activities and formats (e.g. whole group, paired share in breakouts, using chat) also helps engage those with various levels of comfort with public sharing and helps bring more voices into the room.

We have also adapted or created a number of rituals that help form the sacred container for this work in the virtual format. One example is the ritual of what we call "waving each other in" in which everyone on screen is invited to wave to one another so that we remember that we are connected to real humans at the other side of the screen. Additionally, we developed a ritual of blessing hands after people read their Witnesses. Rather than commenting or giving thanks in the chat, in this ritual folks put their hands up to the screen after someone has shared as a way of both saying "I have heard you" and as a gesture reminiscent of the Priestly Blessing in Jewish tradition, offering care and protection to that person. It is powerful to see dozens of people holding up their hands to the screen after you have shared something deep or vulnerable and to be held by the community in this way. At the end of art-making we invite an optional gallery view in which folks who would like to hold their work up to the camera all at once so that we can, with soft eyes, look at the gallery of images that have emerged during the session and remember that each of us are connected to one another through the Creative Source, which moves uniquely through each of us. We end programs with the ritual of sounding the singing bowl, with the intention of letting whatever has been shared that we don't need to hold on to float away on the tone, and whatever we do want to hold onto settle into our bodies. While each of these have a resonance with aspects of Jewish text and tradition, they can easily be translated into any learning environment and adapted to any physical capabilities.

In creating and evolving the ritual practices of this method, we are conscious of the tension that often exists between fixed practices and space for the necessary unpredictability of the creative process to manifest. Navigating this spectrum is the heart and soul of the Jewish Studio Process. Our design draws on the power of ritual to train us to become more adept at welcoming and working with the unpredictability that arises. The purpose of ritual is to provide a set container through which something profound, new and raw can come through. We use ritual to provide sufficient predictable aspects that calm the nervous system enough for learners to be able to recognize what's on their creative horizon, which might yet be out of focus or unseen. As they soften and open they are more able to become receptive to what they find. It is from this place that curiosity can be activated rather than resistance or agitation. Any educational setting would benefit from ritual practices beyond simple comfort but to actually help to prepare the learner for challenging learning.

9.9 Digital Technology Learnings

Since March of 2020 we have facilitated hundreds of spiritual and educational programs for people across the country and around the world. We were able to hone, refine and gain from the many benefits that digital technology allows for and discovered myriad new possibilities that emerged from doing this work online. An online format supports accessibility for those who were not able to make it to our physical studio space to be able to participate, including those who live in other locations, those with limited mobility, and those with other disabilities or sensitivities that make being physically in shared group space challenging. Additionally, many people shared that they felt able to be more fully themselves without barriers such as concern about physical self-presentation, other people seeing their creative process up close, or the need to socialize, which can be over-stimulating. Many people feel more able to safely express vulnerability and openness from the safety of their own home rather than being in a public setting. Additionally, in many cases there was a feeling of a flattening of hierarchies between participants and facilitators when everyone is calling in from home. For example, whether participant or facilitator, all deal with similar issues that might interrupt a session such as kids running by or needing to be attended to, dogs barking, etc. This flattening of hierarchies led to a greater sense of compassion and empathy for one another.

We also discovered the benefits of digital technology as it allowed for greater personal agency and autonomy. Participating from one's own space inherently allows for more control over one's experience. Participants have the option of turning their cameras off, physically setting themselves up in whatever way is most comfortable for them, participating from any location including their yard, a park, or even their bed, and eating and drinking and attending to their own personal needs without disrupting the group. In addition, participating from home, depending on one's living situation, can allow for the feeling of "retreat" without retreating. Folks can gain respite within their ordinary lives, which can help with integrating the tools, practices and learning into how one lives day-to-day. Without having to travel to a designated location, participants conserve energy, lessen stress and can experience a more seamless quality of life that can be calming. Calling in from one's home also opened up possibilities around art materials. Rather than JSP providing an abundance of materials, practicing this methodology from home invites each person to use what is on hand, instilling a foundational value of the JSP: that we already have everything we need and inviting people to look at regular household items they have in new and inventive ways (e.g. candle wax, old newspapers, tin foil, spices) and imagine how they might be used in art-making.

9.10 Digital Technology Pragmatics

Over the course of 2020–2021 and adapting our programming to a digital format, our reach has greatly expanded. While the JSP physical studio could accommodate roughly 20–25 people, online we have no limit. We now frequently have between 60–100 people at a given program. Further, because folks can join from anywhere, our community has gone from primarily local, Bay Area-based to national and even international. Because we are not traveling to facilitate all over the country (e.g. no air travel for participants, no air travel for facilitators, no shipping materials, no swathing conference rooms in plastic to protect the walls for art-making) we are having discussions about the ecological implications of work done in conference and other such locations. Additionally, there is less disruption in our family life. We can facilitate programs and then immediately be downstairs helping our kids get dinner or putting them to bed. Even not commuting the mile and a half from our home to the Studio removes a layer of stress and effort.

9.11 Transferring These Practices to Other Educational Settings

As we've all learned to appreciate the gifts and challenges of the digital world as we've adapted during pandemic times, this time has also raised and highlighted problems including screen fatigue, isolation, and amplification of previously overlooked mental health challenges. In addressing these issues, secular learning environments have much to learn from religious and spiritual educational settings about how to tend to the emotional and existential needs that are always present – and either interfering with or supporting – any learning. There are practices that can be extrapolated from the Jewish Studio Process that speak to basic human needs and can be adapted to any setting. For instance, the deep thinking and tradition of chevrutah can contribute much to paired learning in any classroom (Holzer and Kent 2014). The use of rituals and grounding practices to prepare the learner to relax and achieve receptivity are applicable to any learning environment and can support students outside the classroom as well.

In order to use and adapt these and other practices suggested in this chapter with integrity, the educator needs to have reflected deeply on what needs these practices might serve and must tend to these needs and adapt these practices for themselves in their own lives first. From this, one will gain insight on how best to bring these practices to their learning environment. Educational systems can provide these learning and reflection opportunities to their staff to support their ongoing development and to alleviate fatigue, overwhelm and burn out.

9.12 Conclusion

Practicing a form of transformative education in a time of upheaval, both spiritual and cultural, has yielded new learning and is transforming JSP as an organization as well as a technology of change. We are just beginning to understand all the value that digital technology offers to increase our impact, invite greater diversity of community to flourish and to adapt our methods to specific cohorts such as providing spiritual and creative support for social activists, tailored programs for young Jewish professionals whose careers have been upended by COVID-19 and engaging parents whose family and work life disruptions have revealed needs that were invisible before the pandemic. Additionally, we are engaging with how the insights of the *beit midrash* can translate into other educational settings beyond the Jewish world to help transform the polarization of conversations across the political spectrum that threaten to suppress learning in our schools.

We also acknowledge that we continue to learn, along with our whole country, about the fractures and inequities that exist between and among us. All the wonderful things described here only can accrue when your home is safe and secure, your basic needs met. All the more reason why creating these technologies and making them as accessible as possible in as many settings possible is critical in this work.

References

Allen, P. (1995) *Art is a way of knowing.* Boston: Shambhala Publications.
Allen, P. (2005) *Art is a spiritual path.* Boston: Shambhala Publications.
Armstrong, K. (2020) *The lost art of scripture: Rescuing the sacred texts.* NY.: Anchor Books.
Holzer, E with Kent, O. (2014) *A philosophy of havruta.* Boston: Academic Studies Press.
Gendlin, E. (1982) *Focusing.* NY: Bantam Dell.
Naumburg, M. (1966) *Dynamically oriented art therapy.* NY: Grune and Stratton.
Naumburg, M. (1928) *The child and the world.* NY: Harcourt, Brace.
Prinzhorn, H. (1966) *Artistry of the mentally ill.* NY: Springer-Verlag

Rabbi Adina Allen co-Founder & Creative Director of Jewish Studio Project (JSP), is a spiritual leader, writer and educator who works to help people reclaim their creativity as a powerful, relevant, and transformative tool for personal and social change. Drawing together a lifetime of experience in the creative arts with her rabbinic training, Adina has pioneered a methodology for the integration of Jewish learning, spiritual reflection and creative expression that she has brought to thousands of Jewish educators, professionals and lay leaders across the country. Adina grew up in the studio, learning from and creating art alongside her mother and is blessed to collaborate with her on the writing of this chapter.

Pat B. Allen Patricia (Pat) B. Allen, Ph.D., A.T.R., is an author, artist, art therapist, and teacher who connects to the Creative Source through art and writing. Her books – Art Is a Way of Knowing (Shambhala 1995) and Art Is a Spiritual Path (Shambhala 2005) – explore the borders between art, psychology, spirituality, and social action and are considered classics in the field of expressive arts therapy. Pat co-founded the Open Studio Project in Chicago and of Studio Pardes in Oak Park, IL. She is currently Senior Consulting Faculty for the Jewish Studio Project, which was founded by her daughter, Rabbi Adina Allen.

Chapter 10
Creative Pedagogies with Technology: Future Proofing Teaching Training in Music

Leon de Bruin and Bradley Merrick

Abstract In this chapter, the authors will consider the benefits and challenges of enacting creative pedagogical approaches in the tertiary context and examine emerging educational practices about twenty-first century learning and technology. Creativity continues to be a key construct for twenty-first century music education practice and education, incorporating technology that delivers deeper and more profound learning experiences- that paradoxically isolate individual learning yet at the same time provoke reflection, growth, and sustainability. This chapter explores the delivery of a tertiary degree in Music Teaching, specifically addressing the following areas:

- Curriculum design, delivery, and assessment,
- Entrepreneurial approaches to learning through student centred activity,
- Online learning, student access, self-regulation, and self-assessment,
- Learning environments (including online and technology-based practice) that mirror global change, capacities, and expectations.

Through a combination of annotated examples of teaching practice, selected research, and related theoretical reference, this chapter will propose a range of creative, innovative learning solutions. Importantly, this chapter draws on research undertaken with graduate students after their year of learning during the COVID pandemic and subsequently provides insights into these four areas and their influence on the students' learning. This is supported by a discussion of a range of teaching approaches and strategies that can be used to foster creativities and shifts in teaching practice.

L. de Bruin (✉)
Music, University of Melbourne, Melbourne, Australia
e-mail: leon.debruin@unimelb.edu.au

B. Merrick
Melbourne Graduate School of Education, University of Melbourne, Melbourne, Australia
e-mail: brad.merrick@unimelb.edu.au

D. Henriksen, P. Mishra (eds.), *Creative Provocations: Speculations on the Future of Creativity, Technology & Learning*, Creativity Theory and Action in Education 7,
https://doi.org/10.1007/978-3-031-14549-0_10

Keywords Creative pedagogies · Music education · Music technologies · Online music education

10.1 Introduction

Creativity is commonly perceived as critical for twenty-first century learning and teaching where scholarly and popular discourse point to the importance and need for creative thinking skills in learners. Substantial rhetoric focuses on the need to infuse and embed creativity in teaching and learning within education systems (Harris 2016; Runco 2014). Similar to the positioning of creativity, the capacity to use digital technologies is also commonly seen as a core twenty-first century skill. These technologies can be defined to include "any existing or emerging device or tools, the use of hardware and/or software and/or web-based applications in any way to support learning about, the creation of, the sharing of, or the performance of music" (Merrick 2017, p. 171). Incorporating technology can deliver deep and profound learning experiences and it is often argued that the connection between technology and creativity is a key issue for twenty-first century education (Page and Thorsteinsson 2017). Underpinning this is the need to ensure that incorporating technology in music learning is not in itself a panacea; it is the way teachers creatively approach and adapt learning intentions via the use of the technology, i.e., Zoom chats to enhance student wellbeing and communication (de Bruin, 2021b; Johnson and Merrick 2020) and that they are perceived to be worthwhile and easy to use (Waddell and Williamon 2019). The challenges of enacting creativity in education are substantial (Feldman and Benjamin 2006), and it is not uncommon for creativity to be talked about as being enabled by and through technologies. However, enacting creativity with and through digital technologies is both an unclear and underexplored aspect of education (Mishra and Henriksen 2018). In the past year, the adoption and application of new strategies in teaching and delivering learning with technology have been essential to sustain connection, communication, and engagement with learners and teachers across all settings within Higher Education (Merrick and Johnson 2020). This implies that the work of music teachers includes far more than transmitting content knowledge and musical expertise, but rather the differentiation between different kinds of learning; the creative impacts from the social, relational, interdisciplinary, and affective interpersonal strategies that enhance learning (de Bruin 2018b; Pedder et al. 2005).

In this chapter, we reflect on the creative necessity we utilised in responding to delivering a tertiary music education course. This involved negotiating a worldwide COVID-19 event that necessitated mandated State government lockdowns and shifts to online delivery. We begin by discussing and defining creativity, and then consider how creativity intersects with issues of educational technology, followed by consideration of the benefits and challenges to instantiating creativity in education. We explore the adaptive delivery of specific core elements of this pre-service teaching degree, specifically addressing the following areas:

- Curriculum design, delivery, and assessment,
- Entrepreneurial approaches to learning through student centred activity,
- Online learning, student access, self-regulation, and self-assessment,
- Learning environments (including online and technology-based practice) that mirror global change, capacities, and expectations.

We conclude by identifying some broad themes and recommendations for initial teacher education. We assert that creativity continues to be a key construct for twenty-first century music education practice, and that incorporating technology can deliver deeper and more profound learning experiences- that paradoxically isolate individual learning yet at the same time provoke reflection, growth, and sustainability.

10.2 Creativity and Technology in Education

Creativity is a multifaceted and complex human trait that presents a challenge for many educational systems, often bound to standards or norms that seek clear, prescribed or externally consistent framing (Csikszentmihalyi and Wolfe 2014). University cultures can vary dramatically within states as well as across national or international levels and localised cultures and ecologies inform the distinctive ways creativity and creative approaches to education emerge in real-world practices (de Bruin 2018a; Harris and de Bruin 2018).

Most definitions of creativity identify novelty and usefulness as being two key characteristics of creative ideas or solutions (Plucker et al. 2004). Kaufman and Beghetto's (2009) 4C's emphasise the pillars of critical thinking, communication, collaboration, and creativity, placing 'creative metacognition' and contextual knowledge used to make decisions about one's own creative efforts and accomplishments at the forefront of creative impulse. Glăveanu (2013) emphasizes the role of social factors in which creativity exists in tensions between individuals and groups, his 5A model incorporating actor, action, artefact, audience, and affordances, relying on interdependence of these generalist traits.

Creativity can be described as the production of useful solutions to problems, or novel and interesting ideas and artifacts across domains (Amabile 1996; Zhou and George 2003). This emphasises that whilst there may be generic creative traits, there are multiple creativities that reside in and across each domain, and that are particular to music (de Bruin 2016; Haroutunian 2002). Beyond common themes of newness or effectiveness, whole system approaches (Csikszentmihalyi 2014) as well as ecological overviews (Harris and de Bruin 2018b) emphasise how creativity is located as an interaction within a system of actors in education. In viewing creativity as a thinking skill (OECD 2019), creativity more domain-specifically contributes to the development of metacognition, inter- and intra-personal problem solving, and social dynamics particular to music engagement.

This perpetuates the prevailing tension in which music educators recognize the importance of creativity but are unclear if or how it can be facilitated in their studio or classroom. Thus, the problem of concrete implementation of creativity is at odds with the conviction across educational discourse that creative thinking is important (Sawyer 2015). To infuse creativity into education systems, the field must attend to it at multiple systemic levels, including teaching pedagogy, assessment, resourcing, and policy.

There exists much disparity concerning educational technology implementation in education (Mishra et al. 2011). It can be used to maximize affordances for creative output or deep learning, or it can simply be a replacement device with shallow uses for learning. Complex constructs require complex treatment in how they are approached and applied. Technological requirements in education are becoming increasingly situated, and there has been much consideration of what teachers need to know to use technology effectively in the classroom, and the competencies needed to develop digitally fluent, creative students (Mishra and Mehta 2017). This reflects our definitional overview of educational technologies beyond hardware/software-based opportunities, but also the interactions and transactions that can be afforded to users individually and collectively.

Threading our polymorphous notion of creativity through teaching and learning with these technologies, we incorporate creative pedagogies of dialogue and behaviour (Bakhtin 1986). We build on the Kaufman and Beghetto (2009) 4C's approach by locating teacher-student discourse and relationality such as insightfulness, empathy, receptivity of learner needs, and direct responsiveness to the forefront of creative intra-action (de Bruin 2021b). These qualities are used to conceptualize these behaviours and their relationship to each other that is reflective of the idea that creativity is emergent within a system and within a set of pivotal interactions that are reliant on teacher and student interconnectivity and dialogue (de Bruin 2021a).

The effects of globalization and digital advancement in our world continue to impact on how humans now live, work, think, communicate, and create. Digital tools, devices and applications afford a new world of opportunities in which people can imagine, make, and share in creative ways (Zhao 2012). As knowledge bases continue to expand our approach to complexity, so too has our need for creative thinking and intra-active dimensions in education that address twenty-first century problems (de Bruin 2021a, b; Flew 2011; O'Connor 2010).

In relation to teaching in Australia, the standards of professional practice for teachers is set by the Australian Institute for Teaching and School Leadership (AITSL n.d.) in response to a prescribed Australian Curriculum (ACARA 2016) While creativity is not a core standard for teachers or leaders, it is indicated in the standards, as one of the goals of effective teaching strategies, that "proficient" teachers "select and use relevant teaching strategies to develop knowledge, skills, problem solving and critical and creative thinking" (APST 3.3). In addition, school leaders are expected to promote creative, innovative thinking among colleagues. However, within the AITSL standards, creativity (or creative thinking) is not defined. It is embedded as one of several goals for one particular standard, amongst 37 other standards that are intended to comprise the capacities identified as being essential for teachers in the twenty-first Century. Moreover, there is no indication of how it can be identified, measured, or developed, which undermines the point of any system of standards and makes it difficult to enact in practice. As teaching standards continue to be challenged in their definition of creativity in various contexts and environments, international organisations continue to place an increased value on the development of creative thinking amongst all student learners (OECD 2019), further highlighting its importance in developing twenty-first century capacities for music education.

Amid the shifting context of globalization, rapid digital change, and the grappling of how creativity can be enacted in classroom settings, the COVID-19 worldwide pandemic brought these very conditions into stark and vivid reality. Historically, the delivery of a large proportion of teaching has been done through systematised curriculum and assessment structures and processes, whereby the learner has engaged in activities in the same room as others and drawn on the guidance of the teacher in a range of ways. The 2020 pandemic has brought practical, ideological, and philosophical change to education and teaching practice, the ramifications of which will continue to shape the way we cultivate teaching, learning, and understanding for years to come. In practical terms, institutions pivoted to online learning within a matter of days, adjusting the delivery in a wide array of subjects. This has subsequently promoted critique of traditional approaches, rigid boundary lines between subjects, linear assessments, and restrictive practices for students and teachers. The rhetoric towards creativity and technology and even teacher and student adaptation as an essential aspect of institutional survival has been brought to stark focus.

The adoption of, and continual refinement of these changes required both an openness, and a willingness to innovate and trial new thinking, as practitioners and researchers. Critical to this journey was a willingness to continue exploring and redefining new pathways in the creation of curricula, assessment, and course delivery in teaching preparation programs, to foster capacity and confidence amongst participants, while developing and embracing creative modes of day-to-day practice amongst music teachers.

10.3 The Course and Context of the Study

The Master of Music (Performance Teaching) program at The University of Melbourne is best defined as a bespoke instrumental music teaching training program that specifically looks to build on the students' ability as musicians and teachers, providing them with a graduate learning experience that develops their knowledge of teaching philosophies, methodologies, curriculum, assessment, and pedagogy across a diverse range of subjects and settings.

The course is designed to allow students to select from two distinct streams or pathways as they undertake their study, selecting either 'Studio' or 'Instrumental'. If students feel that they wish to work only in the setting of one-to-one music teaching as part of a private home studio or academy, they can study and develop their practice via subjects and teaching placements that specifically relate to this mode of music teaching in the 'Studio Stream'. The second option involves a wide range of teaching opportunities that involve one-on-one and group teaching scenarios, instrumental and vocal pedagogy, engagement in teaching through music ensembles (chamber, jazz, popular, choral, instrumental, bands, etc), and developing capacities that align with work in a school or larger community setting. This 'Instrumental Stream' importantly develops teacher connection and placement within schools, and provides a substantial amount of in school placement days, which can be used towards their full teaching registration, should they wish to work solely in a school or educational setting.

Through the connection of students with an array of expanded teaching pedagogies and a diverse teaching staff within the Music Conservatorium, the students can continue their development as instrumentalists and vocalists, experiencing and learning about teaching from experts in their discipline, many of whom are performers in their own right. The dualistic mode of engaging all students through the process of understanding and exploring a range of specific teaching approaches via coursework offers focused subject choices such as 'Music Performance Curriculum and Assessment', 'Applied Instrumental and Vocal Teaching', 'Music Learning Teaching and Research', 'Performing to Teach', and 'Evidence-Based Performance Teaching'. This allows the students to investigate, experience and apply the latest research into music learning and teaching across a range of personalised and contextual experiences that are both unique and purposeful to the student. Whether the student is a classical pianist from overseas, a local vocalist from music theatre, or an orchestral percussionist and drummer, the course connects with, and enhances the students' musical experiences and teaching knowledge by allowing students to develop and apply learning to their 'lived experience' as musicians. Importantly, through vicarious observation of teaching (both online and face-to-face) as well as utilisation of teaching frameworks that embrace of processes of self-reflection and self-regulation, students develop confidence in their ability to meet the demands of music teaching when they complete their degree.

10.4 2020 COVID Context

Instrumental music and music teacher education tuition in Australia for much of 2020–21 was met with considerable challenge on the part of both teachers and students.

Changes in teaching and learning in the State of Victoria and Australia saw the University of Melbourne Conservatorium of Music enact adaptive adjustment and development as the COVID-19 epidemic forced governments to mandate societal isolation requirements that greatly impacted teacher practice. The state of Victoria, Australia implemented highly restrictive social measures that included six designated lockdowns including mandatory distancing spanning over 200 days, stay at home rules that required Education departments and Universities to logistically prescribe how schools and institutions would operate.

Universities 'pivoted' to online teaching and learning, with workshops, tutorials and lectures delivered via zoom. Secondary school students also returned to learning through the medium of remote access, with schools using various online media platforms to organise and connect with classes as both teachers and students remained at home yet engaged in teaching and learning. This posed challenges for both our initial teacher education students, their mentor teachers, and students, each of which engaged with each other despite being in the confines of their respective homes.

10.5 The Study

Utilising constructivist principles that forefront students' construction, understanding and experience of learning (Agres et al. 2016), this study explores how instrumental music students in the Conservatorium of Music of Music (Performance Teaching) course engaged and were supported in their learning despite the necessity of activity solely via online platforms between March and December of 2020. The research specifically asked, what practices and behaviours facilitated connection and positive learning engagement in online delivery and adaptive teacher pedagogies? This heightened engagement and increased focus on the completion of student work via online modes (synchronous and asynchronous) challenged teacher pedagogy, student motivation, goal setting, and maintaining of teacher-student connectivity throughout the year.

Participants responded to an online questionnaire that prompted open-ended, written responses to the questions posed. The items specifically asked the students about the situated (COVID-19) experience and included a range of questions such as: During COVID-19, what aspects of the online experience have you enjoyed and how have these opportunities assisted you to develop as a learner? What were the most beneficial/difficult processes in learning as COVID-19 scenarios played out within the course? What were some of the creative teaching approaches (requiring adaptability, flexibility, novel, and critical thinking) that you found most beneficial during the COVID-19 online experience and why were these valuable for you? The study also included a teachers' reflective component, whereby the instructors in the course also responded to these questions. The data provided the opportunity for comparison and analysis between student and teacher experiences, which are discussed in the findings.

10.6 Findings

Analysis of the findings gathered two significant themes emergent across both student and staff data: Confidence in technology, and Cohort connectivity. Staff identified a range of different ways in which they adapted delivery in the course through creative modes and change.

10.6.1 Confidence in Technology

Staff considered the regular uploading of slides and a recorded screencast or podcast before the class for that week. This enabled students to immerse in learning prior to classes, reading and viewing content in advance, which assisted the review of work combined with student thought and questioning. Staff felt this process facilitated depth of learning and fostered more in-depth discussions. Student's feedback was a positive preference for maintaining this model of learning delivery. Staff indicated they were able to support modified online learning activities by combining both the synchronous and asynchronous delivery modes for the students, which increased accessibility for the learner. This was achieved through focused task design and lesson activities that enabled both development and integration of more intense and sustained engagement amongst students as they worked and collaborated online. This was also achieved via diverse modes of delivery and submission, including audio, video, written tasks, blogs, demonstrations online and traditional submissions, all of which facilitated students in modifying their approach to learning tasks and demonstrating their understanding in ways that were both meaningful and personalised. Teachers felt this was an important consideration during this period of restricted access and lockdown.

Both teachers and students discovered that new ways of engaging with learning, discussions, and teaching students were key to making online studies an impactful and efficient learning event. Underpinning this was the need to sustain communication, clarity, and community within the learning environment. As teachers became more adept with zoom functions such as 'waiting rooms' and 'breakout rooms' they were able to adapt their pedagogies, using more sophisticated and diverse strategies to make classes, learning activities and the general dynamic of the online learning experience more diverse and interesting, while also creating purposeful interactions by allocating students with learners of similar interests and experience. Waiting rooms were used for staggered assessments of individuals and groups, and breakout rooms were used both as a strategic pedagogy to connect specific learners in varied locations, and with a degree of improvisational flair to adapt to various learning scenarios. Students were also able to present their learning individually, collectively via a myriad of modes (audio, video, blog, discussion etc) to cater for the diversity of learners. If discussions became energetic, teachers were able to use breakouts to go deeply into student thinking and reflection, and then share various ideas from groups as they emerged. Students commented on their learning:

> I think teachers adapted over the period of time and integrated more approaches, such as using breakouts- I got to know all my fellow students this way- and that would never happen under normal circumstances. We also viewed performance videos and discussed them the next day. That allowed for much deeper and serious questioning of learning and performance. When students demonstrated techniques, the camera was right on the hands- it provided a much clearer demonstration than if we were spread out in a classroom.

Engaging online provided a much more varied and improvisational aspect to classes. I think teachers adapted over the period of time and integrated more approaches, such as viewing performance videos and discussing them the next day. That allowed for much deeper and serious questioning of learning and performance. When students demonstrated techniques the zoom camera was right on the hands- it provided a much clearer demonstration than spread out in a classroom.

The overview of various initiatives students identified as being impactful cover a diverse range of specific tasks, opportunities and strategies that highlight the adaptive, creative modes of teaching that evolved. Students expressed gratitude in being able to submit their music performance (Practical examinations) and teaching demonstrations via video and online upload with many students identifying that this opportunity allowed students to feel confident and tech-savvy by partnering with the teacher in this space. Students expressed that the use of flexible tasks as assessment items that allowed students to engage purposefully and professionally with teaching peers. As one student remarked:

> All of the written assignments completed related somehow to my teaching career in some way. Whether it be reflecting on going for a teaching position (writing a philosophy of teaching assignment) or trying to create a program for a senior secondary instrumental student (creating a VCE performance program), all these assignments were really helpful and gave me confidence and purpose.

Another student reflected on their growing confidence:

> The experience has made me more independent and self-motivated, and confident in my use of technology. It's been positive and I have grown as a person. I embraced change in learning given the pandemic, and I think I adapt fairly well. My skills were amplified through the online environment, in developing tech skills and responding to my studies in new ways.

10.6.2 Cohort Connectivity

Both staff and students felt that despite isolation and distanced learning, qualities of connectivity, and community were nurtured and maintained during the pandemic learning and that this was a significant positive influence on learning. Staff suggested that a range of comprehensive 'Blended' teaching approaches such as pre-recorded lectures, screencasts, and podcasts (asynchronous/flipped), combined with the opportunity to prepare student breakout ideas and activities ahead of time, was an enabler of student connectivity. Ongoing peer interaction provided reinforcement of key topics and learning outcomes from the course curriculum. Teacher pedagogy focused on the opportunity of students to engage with peers, allowing the students to expand their knowledge of topics.

This was further nurtured via online 'in-class' collaborative tasks such as the Zoom whiteboard – whereby teachers used annotations, and discussion threads on Canvas (LMS). These strategies were effective in that they facilitated interactivity, receptivity and communication between students which helped to negate the loss of face-to-face discussion. In fact, discussion became more critical and brought to the fore specific creative music learning processes and the discussion of ideative, iterative and prototyping of thinking and actions in a communal forum. Teachers adopted purposefully designed these fora to facilitate this investigative and disclosing group instruction that enhanced the opportunity to experience diverse and creative teaching strategies and allowed students to experience them first-hand.

Staff initialised ways of enhancing learning online by means of a Zoom Café (an informal meeting of multiple students via Zoom). This provided an opportunity to develop and maintain empathic mentoring relationships with students through the online mode and create a relaxed, informal mode of shared opportunity with all involved. Through the use of relaxed conversations via Zoom, the students identified they were able to catch up on information, find out about subject updates, while also have the opportunity to ask questions about the course and check-in about curriculum or assessment requirements that may not have occurred in a regular subject class.

Students found innovative online modules, such as a newly developed 'Video Resource Hub' was a successful means of engaging students in 'readings' as well as cultivating senses of connection and community. The Video resource Hub (VRH) was housed on the main LMS platform, which allowed students to contribute and view a range of performances, teaching episodes and learning experiences. Importantly, students could revisit these regularly and find materials relative to their own instrumental/vocal and teaching experience. As one student said:

> The video resource room was particularly engaging and allowed me to experience learnings from instrument families other than my own. It allowed me to share ideas and learn from others in my class and across the whole course.

Another student remarked of the responsive and empathic way teachers-maintained connection throughout the 7-month isolation period:

> The faculty have always placed a strong emphasis on connection and of students and of staff looking after students, and of students looking out for each other. This made the online setting a social event that I began to look forward to each week.

Teachers perceived that isolation for an extended period would unsettle and alienate some students from the coursework, sociality and reciprocity that are at the heart of dialogic communication and collaboration. Teacher connectivity to student learning and welfare was a noted aspect of student acknowledgement of staff and a significant part of the ways in which teachers helped students stay attuned and on track. Students remarked that:

> I missed the meeting of friends and gatherings before and after class-we would discuss ideas in a relaxed atmosphere. This was really taken away and attending and concentrating on content online was risky as I always was anxious that I would not understand something and have to hold the class up by interjecting, or that connectivity would cause dropouts – which happened occasionally. Meeting informally in the online Café allowed us to discuss issues with friends and staff, and that was a big help in making me feel on top and connected to my studies.

10.6.3 Discussion and Conclusion

Students felt learning online, though an evolutionary process provided enlightening and perhaps unique aspects to the quality of teaching and learning. Despite initial trepidation, for many, a sense of success and increasing motivation was enhanced through multi-faceted and adaptive approaches to online learning. These approaches enabled access to materials via different technologies (essential for students not in the local time-zone, and with restricted access to the course due to inter-state or international distancing created due to COVID).

Some students identified challenges of being confined to an online environment. Others identified the challenge of learning online and not being able to make music 'on the stage' was a considerable challenge, with one student saying this was "the most difficult aspect was not having the opportunity to be engaged in any music-making. I am definitely the type of person who thrives on musical, academic and social interaction in ensembles and performance".

Constraint can be a powerful promotor of creative action, and negotiating COVID adjustments compelled students to individually and collectively adapt through incorporating technology in generic ways, but also in the calibration of usage in a highly personal and at times intricate ways. Students changed their views towards online formats of learning and acknowledged the immense value of informal online exchanges, whereby they could learn about the course and have exchanges with others as part of a broader Zoom chat conversation that also maintained connection and student community. Most notable were the organisation of the Zoom Cafe cohort meetings. This was such a great opportunity to learn about other students in a forum in which they could listen to others and share information with peers and staff- sharing personalised creative approaches and reinterpreting and transforming ideas for their own personal use.

Yet, many students noted frustration at the online scenario. Students identified that they enjoyed the positive opportunities to break-out in online sessions and talk and interact with their fellow colleagues, although at times they also identified that this was not the experience they had hoped for in their study, feeling that they had missed out on a 'proper' university networking experience, which was traditionally facilitated through face-to-face interaction.

Teachers found an interesting troubleshooting dynamic evolved across the seven-month isolation. They found that students tended to spend more time investigating, developing, and completing their projects than in comparable, conventional courses- perhaps as students were held captive in their homes there were less distractions. Zoom Café meetings allowed troubleshooting questions and discussions to occur outside valuable class time at a designated point in the week, and this made online delivery more efficient and purposeful. With the culture of breakouts being used, learning became more student-centred which was often enabled by their growing confidence in speaking and interacting with each other online. The benefit was that students learned to moderate engagement and discussion as well as manage and distribute their work and ideas in small teams.

A notable feature of the shift in delivery was that in comparing data to previous years, absenteeism showed that more students attended the online lectures and fewer students were absent from the course compared with conventional teaching. All teachers felt that the online approach was more time consuming than a conventional face to face course and required more advanced preparation, combined with sustained communication throughout the term. Lecturers needed a variety of skills beyond those of basic lecturing. They were required to lead student groups towards learning goals and connect the purpose of learning, activating students, facilitating discussions, and assisting students to visualize results from problem-solving processes. Lecturers often led student learning to grappling with how to employ the various media formats and technologies from a user point of view (Merrick 2020). Lecturers also reflected upon teacher-student actions and behaviours, developing the capacity to know when and how to shift between their multiple roles transparently. This from a creative pedagogical standpoint required calibration and differentiated approaches to 'coaching' and 'scaffolding' learning, as well as providing a relationality and cognitive apprenticeship through their roles and responsibilities, and orienting students to new ways of thinking and adapting (de Bruin 2019).

Despite the challenges and adaptations, there were myriad and perhaps unforeseen benefits to this extraordinary online learning event. Online oriented learning made for a rich choice of online material that was made available quite easily. Online learning and teaching facilitated good interpersonal relationships between students and teachers. It brought to the fore more relational aspects of engagement in which teachers and students were compelled to unite with a social countenance of empathy, responsiveness, and insight in understanding of each other (de Bruin 2021a, b). Online learning availed students the opportunity to learn from multiple examples and perspectives rather than just from a single (teacher) one. With appropriate pedagogical tailoring, specific musical creativities and learning was more feasible and mixed modes of evaluation including self- peer- and instructor evaluation was quite easy to adopt.

Online learning provided a medium in which the relational and pedagogical choices allowed learners to be more active, taking on different roles such as breakout group 'leader', teammate, enquirer, evaluator, or recorder. Online learning fostered a higher level of connection, dialogue and respect for each other's opinions that facilitated a more open, cooperative, constructive, and responsible learning environment. Students who tended to be quiet and less expressive in face-to-face discussion seemed to participate more actively in online activities especially when they could consider their responses. These processes challenged student thinking, developing the individual needs of learners while fostering their sense of self-regulation, drawing on features identified in the Adaptive Teaching Framework for Teaching Music Online (Merrick 2020), which draws on twentieth century capacities, evolving teaching roles, learning activators, and curriculum and assessment design.

This poses implications for tertiary education, in utilizing more online teaching through specific design as well as what may be a long-term necessity across many parts of the world. With the teacher not being able to be 'sage on the stage', there is the need for teachers in higher education to adopt the role of facilitator or coach.

Online paradigms require educators to support students, utilizing dialogic pedagogies that engage student thinking and collaboration with each other in online learning activity, as well as illuminating independent thought towards student engagement.

The need for Higher Education teachers to be adaptive and creative with their approaches to curriculum, pedagogy and engagement highlights the "ontology of teacher-student connectivity that promotes an understanding to the precursors, promotors and prevailers of student learning (de Bruin 2021a, b, p. 8). Music teaching, whether one-to-one or in a group is a dynamic and situated event from which creative actions are ignited, evolved and honed. Students are not merely the objects of the teachers' behaviour, they animate and co-construct learning processes with teachers. Sustained online teaching remains a novel approach, that for many educators will require an attentiveness and attunement to engagement that serves to maximise student learning.

Creativity has become a core issue for twenty-first century teaching and learning, and the field of education is one that is still understanding what creativity means in practice and policy. The circumstances of the 2020 pandemic have brought to stark reality that teaching in and of itself happens in a complex space, where there is need for flexibility, adaptability, and support for creative awareness and practice. The notions of insightfulness, empathy and responsiveness remain ever potent in their capacity to cultivate positive behaviours that support learning relationships. Lastly, the need for institutions to be more adaptive and responsive to evolving modes of delivery, engagement and assessment of students remains significant.

Institutions should rely on their adaptivity both within and across faculties in cultivating ecological approaches. Rather than top down and single loop direction, institutions may consider integrating thinking and actions in which successes and challenges are shared more dynamically across faculties and teachers. Employing such double-loop processes to evaluating teaching and learning can instigate innovative and sustainable development and change in course structures that promote creativity and impact transformative educational change beyond the current COVID-19 scenario (de Bruin 2018a, b). We have learned much in our initial response to the unprecedented social events and how this impacted teaching and learning across the globe. These lessons offer insights into how creativity in our future organization, adaptability and preparedness can pave new ways forward in teaching and learning.

References

Agres, K., Forth, J., & Wiggins, G. A. (2016). Evaluation of musical creativity and musical metacreation systems. *Computers in Entertainment (CIE), 14*(3), 1–33.

Australian Curriculum and Reporting Authority. *Australian Curriculum 2016*: *The Arts F-10.* Commonwealth of Australia.

Amabile, T. M. (1996). Creativity in context. Harper Collins Publishers.

M.M. Bakhtin. (1986). Speech Genres and Other Late Essays. University of Texas Press.

Csikszentmihalyi, M. (2014). Society, culture, and person: A systems view of creativity. In *The systems model of creativity* (pp. 47–61). Dordrecht.
Csikszentmihalyi, M., & Wolfe, R. (2014). New conceptions and research approaches to creativity: Implications of a systems perspective for creativity in education. In M. Csikszentmihalyi (Ed.). The systems model of creativity (pp. 161–184). Springer.
de Bruin, L. R. (2016). Theory and practice in idea generation and creativity in Jazz improvisation. *Australian Journal of Music Education*, (2), 91.
de Bruin, L. R. (2018a). Organisational Change for Creativity in Education. In *Creativity Policy, Partnerships and Practice in Education* (pp. 167–191). Palgrave Macmillan.
de Bruin, L. R. (2018b). Shaping interpersonal learning in the jazz improvisation lesson: Observing a dynamic systems approach. *International Journal of Music Education. 36*(2) 160–181.
de Bruin, L. R. (2019). The use of cognitive apprenticeship in the learning and teaching of improvisation: Teacher and student perspectives. *Research Studies in Music Education. 41*(3): 261–279. https://doi.org/10.1177/1321103X18773110
de Bruin, L. R. (2021a). Instrumental Music Educators in a COVID landscape: a reassertion of relationality and connection in teaching practice. *Frontiers in Psychology.* https://doi.org/10.3389/fpsyg.2020.624717.
de Bruin, L. R. (2021b). Instrumental Music Education in the time of COVID: A Reassertion of Relationality and Connection in Teaching Practice. *Journal of Music, Health, and Wellbeing*. (In press).
Feldman, D. H., & Benjamin, A. C. (2006). Creativity and education: An American retrospective. Cambridge Journal of Education, 36(3), 319–336.
Flew, T. (2011). *The creative industries: Culture and policy*. Sage.
Glăveanu, V. P. (2013). Rewriting the language of creativity: The Five A's framework. *Review of General Psychology, 17*(1), 69–81.
Haroutounian, J. (2002). *Kindling the spark: Recognising and developing musical talent.* Oxford University Press.
Harris, A. (2016). *Creativity and education*. Springer.
Harris, A., & de Bruin, L. R. (2018). Secondary school creativity, teacher practice and STEAM education: An international study. *Journal of Educational Change, 19*(2), 153–179.
Harris, A., & de Bruin, L. (2018b). An international study of creative pedagogies in practice in secondary schools: Toward a creative ecology. *Journal of Curriculum and Pedagogy, 15*(2), 215-235.
Johnson, C., & Merrick, B. (2020). Enabling music students' well-being through regular Zoom cohort chats during the COVID-19 crises. In R. E. Ferdig, E. Baumgartner, R. Hartshorne, R. Kaplan-Rakowski, & C. Mouza (Eds.), *Teaching, technology, and teacher education during the COVID-19 pandemic: Stories from the field* (pp. 261–264). Association for the Advancement of Computing in Education (AACE).
Kaufman, J. C., & Beghetto, R. A. (2009). Beyond big and little: The four c model of creativity. *Review of general psychology,* vol. 13, no. 1, pp. 1–12.
Merrick, B. (2017). Popular music and technology in the secondary school. *The Routledge companion to music, technology, and education*, 171–180.
Merrick, B. (2020), 'Changing mindset, perceptions, learning, and tradition: An "Adaptive Teaching Framework" for teaching music online', *International Journal on Innovations in Online Education, 4*:2, pp. 1 17. https://doi.org/10.1615/IntJlInnovOnlineEdu.2020035150
Merrick, B., & Johnson, C. (2020), '*Teaching Music Online in Higher Education:* 2020 conference report', *Journal of Music, Technology & Education,* 13:1, pp. 95–108, https://doi.org/10.1386/jmte_00018_1
Mishra, P., & Henriksen, D. (2018). *Creativity and technology: Rethinking their role in education.* Springer.
Mishra, P., Koehler, M. J., & Henriksen, D. (2011). The 7 transdisciplinary habits of mind: Extending the TPACK framework towards 21st century learning. *Educational Technology, 51* (2), 22–28.

Mishra, P., & Mehta, R. (2017). What we educators get wrong about 21st-century learning: Results of a survey. Journal of Digital Learning in Teacher Education, 33(1), 6–19.
O'Connor, J. (2010). *The cultural and creative industries: a literature review*. Creativity, Culture and Education.
Organisation for Economic Co-operation and Development (OECD). (2019). *PISA 2021 creative thinking framework*. https://www.oecd.org/pisa/publications/PISA-2021-creative-thinking-framework.pdf
Page, T., & Thorsteinsson, G. (2017). The impact of conventional school education on students' creativity. *I-Manager's Journal on School Educational Technology, 13*(1), 12.
Pedder, D., James, M., & MacBeath, J. (2005). How teachers value and practise professional learning. *Research papers in Education, 20*(3), 209–243.
Plucker, J. A., Beghetto, R. A., & Dow, G. T. (2004). Why isn't creativity more important to educational psychologists? Potentials, pitfalls, and future directions in creativity research. *Educational Psychologist, 39*(2), 83–96.
Runco, M. A. (2014). *Creativity: Theories and themes: Research, development, and practice.* Elsevier.
Sawyer, K. (2015). A call to action: The challenges of creative teaching and learning. *Teachers College Record, 117*(10), n10.
Waddell, G. & Williamon, A. (2019) Technology Use and Attitudes in Music Learning. *Front. ICT* 6:11. doi: https://doi.org/10.3389/fict.2019.00011
Zhao, Y. (2012). *World class learners: Educating creative and entrepreneurial students.* Corwin Press.
Zhou, J., & George, J. M. (2003). Awakening employee creativity: The role of leader emotional intelligence. *The Leadership Quarterly, 14*(4-5), 545–568.

Dr. Leon de Bruin is Lecturer in Music at the University of Melbourne, Conservatorium of Music, and co-ordinator of the Master of Music Performance Teaching degree (MMPT). As ASME National President he is a staunch advocate for quality music education in Australia, and music teacher education wordwide as an executive of ISME IVMTC (Instrumental and Vocal Teaching Commission). He has published over 50 articles, chapters, and edited books, including Music Ecologies: Instrumental Ensembles around the World, Revolutions in Music Education, and Creativities in Arts Education, Research and Practice: International Perspectives for the Future of Learning and Teaching.

Dr. Brad Merrick is Senior Lecturer (Music and the Arts Education) at Melbourne Graduate School of Education, the University of Melbourne. Brad completed his PhD in Music Education at the University of New South Wales. He is a Past National President of the Australian Society for Music Education (ASME) and currently serves as Chair of ISME's MISTEC (Music in Schools and Teacher Education) commission. He is actively involved in curriculum reform and advocacy, having published widely, and presented in many different settings, to both students and teachers.

Chapter 11
Why We Should Take a Second Look at the Politics of Creativity: The Dangers of a Celebratory Mode

Shakuntala Banaji

Abstract At the heart of this chapter is an ethical contention: By avoiding scrutiny of creativity sustaining processes, strategies and products serving authoritarian, violent or discriminatory practices, we fail to face the ethical question of what happens when we fetishize 'creativity' either in the abstract or in particular circumstances (such as in regard to digital culture, AI or education) without attending to the politics and effects of its deployment. Leading up to this contention, the chapter draws on evidence from more than two decades of work with children and young people in regard to media use, political or civic participation and contributions to everyday social reproduction to describe a range of political and social creativity. The chapter theorises the way children and young people apply creative learning – and technologies old and new – to everyday survival, politics and activist struggle. Generated from a range of qualitative methodological fieldwork carried out between 2007 and 2020, including in-depth face-to-face interviewing, ethnographic observation, textual analysis and contextual, historical analysis, three youth-centred vignettes at the heart of the chapter offer a necessary provocation around unreflexive normativity when theorising creativity and learning, problematising the non-recognition of *forms of creativity* that do not line up with normative imperatives and frameworks.

Keywords Creativity and discrimination · Global south · Sustainability · Dangerous creativity · Children and youth · Unreflexive normativity

---

S. Banaji (✉)
Department of Media and Communication, London School of Economics and Political Science, London, UK
e-mail: s.banaji@lse.ac.uk

D. Henriksen, P. Mishra (eds.), *Creative Provocations: Speculations on the Future of Creativity, Technology & Learning*, Creativity Theory and Action in Education 7,
https://doi.org/10.1007/978-3-031-14549-0_11

11.1 Introduction

This chapter draws on evidence and experiences from more than two decades of work with children and young people in regard to media use, political or civic participation and contributions to everyday social reproduction. Some of this has been meticulously documented (Banaji and Buckingham 2013; Banaji 2017; Banaji et al. 2018; Banaji and Mejias 2020); some remains speculative. The opening sections of the chapter describe a range of political and social creativity engaged in by children and young people in the course of their daily lives and the way they apply creative learning – and technologies old and new – to survival, politics and activist struggle. The brief vignettes and case studies at the heart of the chapter were generated from a range of qualitative methodological fieldwork including in-depth face-to-face interviewing, ethnographic observation, textual analysis and contextual, historical analysis. The latter parts, in line with the overall aim of the chapter, offer a necessary provocation around unreflexive normativity when theorising creativity and learning. The discussion aims to problematise the non-recognition of *forms of creativity* that do not line up with normative pro-social imperatives. In the conclusion I suggest that by tending to refuse to discuss or engage with as 'creative' processes, strategies and products which serve authoritarianism, violence or discrimination, we fail to face the ethical question of what happens when we fetishize 'creativity' either in the abstract or in particular circumstances (such as in regard to digital culture, AI or education) without attending to the politics and effects of its deployment.

11.2 Creativity, Technology and Lived Experience: Learning to Survive, Teaching to Resist

While the definition of each of these key terms will emerge more fully as we proceed, at the outset it is worth noting that I use technology here in its most capacious sense to denote all material systems, assemblages, artefacts and tools deployed by humans to communicate, support and scaffold life and human dignity, and this is important to note, to undermine and destroy the life and human dignity of some groups while privileging others. There is often an unspoken assumption in the fields of media and education that we all understand *technology* as information and communication technologies or new digital media, and that these usually denote social progress. Meanwhile, whether we think of creativity in this chapter as interlinked processes involving imagination, playfulness, problem-solving and potentially novel assemblages made by individuals and groups, or as some combination of processes and products is less important than acknowledging the unspoken *rhetorics* which cling to this word in our usage. That creativity is conceptualised in different and even contradictory ways within different epistemological, cultural and political traditions, and within different fields, is significant to the argument of this chapter. *Learning* is likewise a term with multiple formal definitions in the field of education but which we will deploy in this chapter as the acquisition of any set of habits, preferences, values, skills, knowledge, understanding or insight without a necessary bias towards learning as a value in itself, since some learning can be faulty, some can be counterproductive and some can be linked to survival but also profoundly traumatic.

So, let us start by assuming that we all know what creativity is – namely that both processes and products are familiar and desirable, that they involve innovation, imagination, constructive problem solving, an orientation towards social good, and so on and so forth. Partly to lay out the connections between this benign version of creativity (which is the one most of us use in everyday parlance), and learning as the generally positive acquisition of helpful knowledge and skills while also interrogating our obsession with the newest of digital technologies in the context of lived experience, I want to begin by recounting some of the different creative processes and products (as well as the circumstances) I encountered over a decade of researching children and media in India. These range from the young domestic servant in a tower block who fashioned a device out of spoons and string to retrieve something that one of her employers' children had dropped off the balcony to the innovative repurposing of a variety of rubber tyres and plastic bottles by teenagers into play-things for their baby brothers and sisters as well as the "jugaad" (जुगाड़) or constant searching for ways of hacking into, altering settings or eluding surveillance by parents, relatives, husbands and boyfriends which seemed to be a common-place of the new digital era.

As I discuss in chapter 6 of *Children and Media in India* (2017: 151–186), I met Mansi and Dheeraj, brother and sister, 8 and 12 years old, both garbage sorters in a rundown suburban area of Bombay. Mansi and Dheeraj's parents, working class contributors to the Indian informal economy, come from Bihar in the north of India; but Mansi and Dheeraj had never been there. They walked around barefoot, lived on the pavement and cooked on a kerosene stove. I never met their mother, a rag picker, or father, an unqualified mechanic. At the time of my fieldwork, Dheeraj was responsible for collecting bits of metal from the garbage along with batteries for their mother to dispose of or sell. While doing this he also came across multiple broken but coveted objects including tube lights and phones which he disaggregated and gave to his mother to be sold as scrap or re-assembled to make tiny non-mechanised objects for his younger siblings to play with. One of Dheeraj's favourite sayings (which he repeated at least ten times during our conversations) was, '*we* [the poor] can live because *they* [the rich] throw things away. Everything can be used somehow, you just have to work it out'. It was obvious to me that in Dheeraj's life, disaggregation and disassembly – both highly creative processes – played as central a role as assemblage; and both assembling and disassembling things allowed him considerable scope for creative agency and learning. You try things, you make mistakes, you fail, you try again in a more inventive way, or a way that you had not considered previously. Mansi likewise seemed to be following in her family's footsteps both in occupation and creativity. Excluded almost entirely from the glittering realm of the digital but nevertheless linked into the digitisation of everyday life through the digital and electronic detritus collected via their family's waste work, Mansi and Dheeraj demonstrated a repeated *creative conservation* that puts billion dollar corporations and million pound development projects to shame.

Varsha was 14 and working for an upper middle class family in a north Bombay high-rise when I met with her to observe and discuss her routines and media consumption over the course of several weeks. We established quite early on that Varsha was incredibly dextrous and that she had a penchant for digital technology – she wanted ardently to 'learn computers' and possess 'a mobile phone' of her very own but knew well that she would be punished for attempting to touch anything belonging to the family she worked for. I gradually ascertained that although her employers had taken her away from her family in order to send her to school and had promised only to ask her to work "a few hours a day", she was not being educated formally and was instead at the beck and call of her employers and their twin 9 year olds almost 18 to 19 hours a day. She was insulted by her female employer, slapped by the children and yet retained a sense of dignified resistance. This was evident in every aspect of her being from her class solidarity with her fellow servant Dhavan, her compassion for her employers' children which would not allow her to let them get punished by their parents, to her constant and repeated creative fixes to everyday problems. As I recount in the book, my observations with Varsha were filled with examples of her creativity:

> On one occasion I was playing with the children and Mickey [the boy] broke a toy helicopter that he had just been given by some 'foreign' guests. Dhavan guffawed, saying 'now he'll get what he deserves'; Mickey started to cry. I did not intervene. Varsha was aware that he would be punished by his father for this transgression: she spent half an hour carefully repairing the mechanism, connecting wires and gluing the plastic so that no adult would know the toy had been damaged even if it did not work. On another occasion Minnie [the girl] was throwing a tantrum and managed to fling her mother's silk shawl off the balcony of their penthouse and it lodged precariously on a metal outcrop used for hanging flower pots and trailing plants. Dhavan was a witness, as was I, so Varsha could not be blamed. However, she took pity on Minnie, rigged up an elaborate mechanism – an assemblage – consisting of rope, a curved cooking utensil, and sticky tape, to reach out and dislodge the shawl. She took time out of her busy routine – exposing herself to the risk of being beaten for time-wasting – to descend to the courtyard of the building to collect the shawl and rewash it (Banaji 2017: 80).

Without romanticising the dire poverty, violence, discrimination and exploitation that working class children like Dheeraj, Mansi and Varsha experience every day of their lives, it is also important to draw attention to the enormous erasure that takes place when discussions of creativity and learning focus only on *digital technologies* limited to mobile phones, platforms and apps while abandoning all the other tools and technologies to some supposedly pre-modern realm. Of course, there was no dearth of digital-aided creativity on display amongst the children and teenagers I interviewed during those years of fieldwork – from computer-assisted musical compositions, programming apps, memes, GIFs and short videos of themselves and their friends performing to digital art and scientific inventions. I was shown the products repeatedly, and I witnessed the processes frequently. There was also an in-between realm of low-tech digital tools such as older mobile phones with no internet connections being touted for their relevance in education and learning as well as other media from television and films to songs and comics that featured regularly in children's play and adolescents' imaginative identity work. However, it was also obvious that amongst the scholarly communities in media and communications, education and international development, an internalised rhetoric of digital creativity was playing a powerful role in preventing other forms of creative praxis from being recognised and explored. Ergo, the deep pedagogic mechanisms of trial and error with multiple materials, tools and technologies deployed by working class children, young people and their families in everyday survival situations, far from being linked to the kind of creativity desired in silicon valley or Scandinavian school curricula, were associated at worst with victimhood and at best with the dubious psychological concept of resilience.

Before turning to the more critical aspect of this discussion, I want to dwell briefly on the varieties of political creativity – or rather creativity applied in the service of political and social change - that I and my colleagues encountered on a cross-national European project, CATCH-EyoU, which investigated young people's political activism across eight different EU countries: The Czech Republic, Estonia, Germany, Greece, Italy, Portugal, Sweden and the UK. A central issue for our project was that of creativity and innovation in active citizenship. Therefore, amongst other questions about representation, voice and access, we asked what, if any, were the most creative and innovative aspects of the selected groups, networks, organisations and initiatives and proceeded to examine how these aspects of innovation and creativity fitted with *traditional* (voting-centric) political participation and *dutiful* (compliant to the state) citizenship?

One of our key findings was that not only were innovation and creativity conceptualised differently by organisations, but also this divergence in definitions also held good for our cohort of researchers. Nevertheless, there emerged several points of agreement. Most of the initiatives examined, including formal and informal organisations and groups, loose political and civic networks, purely online or purely offline youth bodies, and even social movements, actually seemed to rely on a limited range of well-known and accepted activities and strategies such as leaflets, websites, social media campaigns, demonstrations, lobbying, civic networking, informational presentations, and regular meetings to carry out their mandate.

In writing about the work of these organisations and networks I found it extremely helpful to distinguish between three types of creative and innovative practices: (1) where an organisation or initiative itself was *built around a surprising and unexpected theme or topic*; (2) where an organisation or initiative's *practices* demonstrated features of creativity and (3) where the *organisational structure* of the initiative was unusual and creative. Although they occasionally converged, these three types of creativity and innovation did not always go hand in hand. The examples I include below demonstrate the deep interconnection between pedagogy, politics, new and emerging communications technologies and creativity. Creative ways of putting democratic values into practice were evident in several of the initiatives. For example, take Piána na ulici (Pianos on the Street) in the Czech Republic. This group places pianos in diverse public spaces across the country and experiments with other ways of building community cohesion and enlivening public space, such as the placement of chess tables and poetry 'jukeboxes' (including, even, a design device). A video of a policeman playing one of the pianos on the street had gone viral at the time of our research, connecting the offline and the online seamlessly. Such attempts at community building clearly are forms of political creativity in addition to having significant pro-social benefits. In Portugal, our colleagues worked with the environmental campaign group Cidade+ who were at the time putting on spectacular but also deeply pedagogic events with multiple material and digital props to simulate, teach and explain aspects of climate change to a sceptical, dubious or uninterested public (cf. Cruz et al. 2020).

Several of the creative practices used by the young activist citizens we encountered were of course *new media-based*. In the early 2000s when social media was rather newer even in Western Europe, only the most cutting edge organisations were making use of those tools and techniques for communication and engagement. Interactivity was at a premium on youth civic websites (cf. Banaji 2008; Banaji and Buckingham 2013). However, by the time we were conducting our research in 2017–18, some of the strategies which might have been innovative at one point in time, had become commonplace and were used by almost every organisation. This was the case with many of the social media campaigns and pages. Therefore, to have an impact, an organisation or network had to think beyond the domesticated uses of new and emerging technology towards ways of deploying and creating such technology that solved pressing social problems or changed minds. In the UK, the socialist youth campaign group *Momentum* were engaging with a multitude of young tech designers, programmers, film makers and editors to design startling and funny viral videos that drew attention to economic injustice or social repression by the British government, as well as engaging in different forms of campaign training and agit-prop theatre to arouse interest in the election campaign of the then Labour Party under the leadership of Jeremy Corbyn (cf. Macková et al. 2020). One of Momentum's most innovative and impactful projects was *M-app*[1] that was said to be able to revolutionise the way young socialist activists could have their voices heard within the broader party on issues of policy and direction. This technological innovation joined their already widely successful app *My nearest Marginal*[2] which was designed to allow Labour Party supporters and members to campaign in areas which needed just a bit more voter engagement to have a strong chance of changing the local Member of Parliament. In addition we noted their use of an amusing and engaging Tinder bot that engaged would-be daters in political talk before going on to discuss dating preferences.[3] Of course, what all of these young socialist innovators in the UK should probably have considered but could not have allowed to get in the way of their creative energy was the fact that conservative

[1] https://www.wired.co.uk/article/momentum-app-labour-party-conference

[2] https://mynearestmarginal.com

[3] https://www.dazeddigital.com/artsandculture/article/36240/1/meet-the-flirty-tinder-bot-swinging-the-youth-vote-general-election

forces with a far greater amount of capital and funding were waiting in the wings to copy and deploy these sorts of creative ideas in service of the very opposite ideological and political goals.

To conclude this section, then, it is worth noting that the most exciting practices we observed, several combined young people's passions for food, art and music with their groups' sense of the need for social justice and/or resistance to authority while others employed creative strategies to embed democracy from the most basic stages of action planning. There were certainly some which centred new and emerging digital technologies – and we observed how young people had to teach their older peers ways of handling and managing these. However, the brief list below (excerpted from over 50 different observed strategies and practices) demonstrates that digital and non-digital practices were evenly balanced:

- Making political or social oriented graffiti and street art with political messages
- Collecting scrap metal, plastic and other pollutants on beaches to clean the environment and then turning these into street art for the public
- Conducting political meetings which would allow everyone to speak and to draw up moral and political manifestos together without excluding anyone (some of these lasted all night and sometimes even for 2 days, with young members camping out together in open spaces or disused factory buildings)
- Film making clubs and film screening co-operatives on youth social issues
- Philosophical discussion groups which brought communities together (eg. Muslims and Christians; Catholics and Protestants)
- Largescale street food events to highlight particular issues around food waste
- Helping homeless people back into the community through social enterprise – giving them unusual jobs such as 'tourist guide'
- 'Guerrilla gardening' and 'Guerrilla farming' – taking over unused land for permacultures
- Street theatre and spectacular events - (projecting symbols onto official buildings, holding mock trials through the use of digital tools)
- Running alternative media on and offline – magazines, newsletters, radio and television as well as social media content aggregators
- Making and circulating apps that help protestors to avoid surveillance by police and other authorities
- Making proxy web addresses to circumvent censorship of particular issues (e.g. to do with protests against the monarchy, or Muslim Lives Matter)

Further discussion of activism and activist practices can be found in our project Blue Paper 'Defining successful citizen action' (2018).[4] The next (penultimate) section moves away from the pro-social and inspiring examples of working class children deploying creativity for survival and activists using creativity to challenge political elites and the capitalist status quo to engage with a series of areas in which critical thinking, innovation and problem solving, and indeed a blend of these, have been

[4] https://www.catcheyou.eu/download/BP8%20(website%20version).pdf

deployed to create systems and tools which allow users to discriminate against, dispossess, suppress and violate people and communities around the world in ever more efficient and extractive ways. Notably for the argument of this chapter, it is as important to grapple with the ramifications of creativity as always open to inflection by different moral and political perspectives – even destructive ones – rather than as having a positive moral weight in and of itself and thus conferring that moral superiority on any and all creative endeavours.

11.3 The Missing Rhetorics of Creativity?: From Military Strategy and Torture to Neoliberal Exploitation and Authoritarian Disinformation

I am writing at a moment when a US drone strike at Kabul airport sent ostensibly to punish a terrorist organisation did exactly what it was programmed to do and killed more than eleven Afghan children. This is only the tip of the horrific iceberg of Afghan civilians murdered by US drone strikes, British and US armed forces and different factions of Islamic fundamentalists with various explosive devices over the past two decades, and leads me to ponder technologies of death. What do these technologies mean for learning about insurgency and counter-insurgency? How should we acknowledge the notion that each of the these death-dealing groups, militias and their political backers have at points celebrated the creativity behind the prototyping of newer and more deadly machines? Distressing as this may be, and far as it is from the intentions of many educators in classrooms, contemporary education systems including tech apprenticeships and job-training often support people in learning how to create and produce systems (Garrett 2021) and tools of discrimination and death. But let me backup a moment to contextualise this shift in perspective.

While writing *The Rhetorics of Creativity* (Banaji et al. 2006) I and my colleagues debated fiercely about the extent to which creativity should be defined as a *relational sentimental value* tied to normative understandings of beauty, productivity, resistance to authoritarianism and pursuit of pro-democratic political change. The fact that creativity is often automatically assumed to connote something constructive and sought after, I argued, does not guarantee that its products (and processes) are necessarily a universal moral good. On the contrary, and particularly in the arenas of politics, spurred by an interest in ecological destruction, neoliberal capitalism, political authoritarianism and settler colonialism, I was coming to view creativity as potentially *either* regenerative and uplifting *or* destructive. While creativity in many spheres can be either amoral or expressive of sets of normative values, where it is embedded in moral value systems, it has as much of an affinity for hate and violence as it does for construction, love and restoration. Or, put another way, creative individuals and their creations do not necessarily have an affinity for justice: they can contribute both to oppression and liberation, there is no intrinsic link between creative production and fairness.

In this vein – and setting aside for the present all the works of art, poetry and music that have long been held as the epitome of creative genius but whose creators were embedded in the oppressive systems of their day whether these be Nazism, racism, colonialism or misogyny – there is no doubt that creative praxis informs the actions of resistance movements. We have only to look at those who conceived of and participated in setting up the underground railroad, the civil rights movement, the decolonial struggles of the mid-twentieth century, the Palestinian intifadas, the civic justice movement of the women of Shaheen Baag in India and so on and so forth. However, while some would by definition rule out such possibilities, creative forms of invention, violence and suppression are also implicated in slavery and colonisation; in sweatshops, military surveillance and killing technologies, and in co-opting discourses of revolution to sell T-shirts and conservative politics. Pondering this tension between the outcomes and aims of creativity in 2006, I argued that it was important to question the taken-for-granted celebratory and sentimental attachment to the idea of creativity as *implicitly benign* and thus always necessarily *normatively desirable* which was evident across almost every domain where this concept was being interrogated.

It is a matter of history and editing that much of this complex theorising did not make it into our final paper. While we agreed to include a section on creativity as art and resistance, my incipient theorising about the 'creative' nature of particular forms of state and nonstate violence, for instance, remained just a little too far beyond the pale for those of us engaging in arguments to include creativity in the curriculum because of its illuminating and democratic potential for educators and students. Perhaps we shared an anxiety that discussion and description would be taken as a form of flattery and endorsement – that naming aspects of the military-industrial complex as 'creative' might imbue these institutions with some false sense of moral good. Our reticence was even odder, considering the many historical sources praising military strategies and tactics by labelling them creative, regardless of the vast populations suffering and dying in their wake and the many contemporary sources advocating for 'creativity' within professional military ranks.[5] Perhaps we were correct not to open that can of worms at that time, and instead to focus on the important task of interrogating the multiple different definitions of the word.

[5]In a classic example of this, Lowther et al. (2020) writing for Thestrategybridge.org ask: 'Is it possible to incorporate a creative mindset into professional military education curriculum?' and go on to argue 'In our *War on the Rocks* article, "Professional Military Education Needs More Creativity, Not More History," we argued that it is possible, but did not explain how. Here, we offer some insight into developing a creative mindset by first explaining what we mean by creativity and then offering steps for embedding creativity education into a joint professional military education program. Our recommendations are drawn from successful creativity programs the private sector invests in for their own employees, from examples already used across professional military education, and from academic research'.

In the 15 years since we wrote that monograph, the rise of social media and artificial intelligence based technological systems has brought to the fore different aspects of the rhetoric at different moments, particularly with regard to creative affordances, creativity as an economic good and as a social good. Companies founded with inherited or venture capital accrued through the exploitation and extraction of racialized labour, and teams of digital creatives such as Google (now Alphabet), Facebook, Amazon, Alibaba, Reliance-Jio and Uber, have monopolised the public sphere, entrenched their economic and social transcendence and overtaken many older, more established industrial and media conglomerates (Dean 2013; van Dyck 2018; Meng 2018). The roots of these large corporate entities are firmly within the neoliberal capitalist system regardless of their faux creative rhetoric (Ross 2009). In fact, the multiple exploitative facets of the 'gig economy' associated with these platforms are frequently theorised through a link between creativity and self-commodification. However, there are other ways of theorising them which link them to a clear and present danger to human rights, civil rights and democracy (Raji et al. 2020; Banaji and Bhat 2019).

In our work on the use of WhatsApp by far right Hindu nationalist government IT-cells and far right Hindu nationalist vigilantes in India to target Muslims, Christians, Dalits and other minorities by discriminating, intimidating or inciting violence against them, Ram Bhat and I traced the types of digitally literate disinformation distributors who were most likely to be involved at the tech end of the business. These people, we found, were engaged in a variety of tacit and open propagandist tactics that drew on their knowledge of programming and building information systems, as well as of algorithmic detection (of misinformation) and how to evade it. These individuals and groups possessed high degrees of ideological certainty about the correctness of their actions in calling for and committing discrimination and murder against fellow-citizens from minoritized groups, while also possessing a chilling degree of creative and innovative knowledge and attitude when it came to creating loop-holes in the technological solutions proposed by the policy and tech teams in WhatsApp, Facebook and other cross-platform apps. To give an example, if WhatsApp limited the number of forwards or the number of groups that a user could join, it was a matter of days before one of these far right IT cell innovators had worked out how to clone the app, use an old version of the app, or some other plug-in which again enabled messages to be sent to hundreds if not thousands of users, or the user to join more than the upper limit on groups by buying and using multiple cheap sim cards. Meanwhile, an Israeli spyware called Pegasus was being used illegally to surveille and harm dissident activists who used WhatsApp in India.[6] All of this happened despite Facebook's claims to be entirely protecting the privacy of their users via encryption and despite the many far right Indian users uploading what were in effect snuff videos and rape videos to WhatsApp groups with celebratory comments. While none of this is even remotely similar to the creative process

[6]https://www.buzzfeednews.com/article/pranavdixit/an-unidentified-government-spied-on-dissidents-in-india

and production engaged in by a Mozart, a Sembene Ousmane or a Simone Biles, within the rhetoric of technological innovation, this is clearly recognisable as creativity, and given that multiple other users then go on to learn from these procedures, there is also an element of democratic creativity and creative learning – invitation to experiment, to collaborate, to play and to do so in a supportive environment – involved.

Meanwhile, in the world of bots, propaganda, disinformation, misinformation and deep fake scholarship, it is still unusual for scholars of creativity to acknowledge the elements of creativity that make such negative innovations possible. When other scholars do discuss creativity in this field, it is usually because they are aiming to entice corporations to invest in their work, with a view to prototype development and subsequent mass production. The likelihood of this connection between business and research is made all the more likely by the decline in public funding of universities and the corporate sponsorship of university labs or 'creativity hubs'. In a paper entitled 'FakeRetouch: Evading DeepFakes Detection via the Guidance of Deliberate Noise' Huang et al. (who hail from the worlds of tech scholarship and business research) explain that '[t]he novelty and creativity of DeepFake generation techniques have attracted worldwide media attention' (2020: 1), before going on to outline their proposal:

> the FakeRetouch, a framework that takes a hybrid approach of deep image filtering and noise addition to improve the fidelity of GAN-based fake images. Our method effectively reduces and destroys the artifact patterns introduced by existing fake image generation methods, in both spatial and frequency domain. Through removing fake artifacts, our further reconstructed image retains photo-realistic and high-fidelity, and can bypass state-of-the-art DeepFake detection methods as well as DeepFake provenance method (Wang et al. 2020b) (2020: 7).

In the sometimes murky world of AI-based tech-innovation that allows images to be morphed in ways which make them indistinguishable from images of events and persons that really exist or existed, what this group of writers are proposing is not in the least uncommon – a protocol for ensuring that DeepFakes will become ever deeper and less accessible to scrutiny by those desirous of 'fact checking' or ensuring the authenticity of images and other information contained therein. Given that such deep fake videos depicting the left candidates in Brazil consorting with paedophiles or apparently engaging in non-consensual sexual activity were circulated by Jair Bolsonaro's infamous 'Office of Hate' and swung the last elections towards the far right candidate, one can only imagine how much further damage those with neo-Nazi beliefs, extensive capital and a will to power will wreak on public opinion. If we are to learn anything from these examples, it is imperative that we heed Raji et al.'s warning that 'it remains challenging for practitioners to identify the harmful repercussions of their own systems prior to deployment, and, once deployed, emergent issues can become difficult or impossible to trace back to their source' (2020: 1) before pushing ever more complex and powerful, but biased and contaminated AI and machine learning into production and out into the public domain. As has been said, just because we *can* invent an atom bomb, does not mean that we *should*.

To take another deeply troubling instance, the methods of 'torture' and 'technologised disinformation' deployed by military regimes, the CIA, and secret police in multiple different regimes are rarely discussed as examples of 'creativity' *in the scholarly literature on creativity and learning*. However, these techniques fit many of the rhetorical descriptors of this concept. This is not to say that the connection has not been noted. For instance, writing of human rights, Natan Lerner notes that 'human imagination has shown great "creativity" in devising means by which to destroy the capacity of prisoners and suspects – guilty or innocent – to resist, sometimes without even touching their bodies but always at the price of horrendous pain and suffering' (1986: 126). Given how ubiquitous contemporary accounts of torture through the infliction of varieties of physical pain, incarceration, sleep deprivation, water-boarding, electrocution, humiliation and psychological warfare are in fiction and non-fiction alike, and regardless of the political persuasions of those advocating the necessity of these vile practices, the connections between power and creativity are apparent. In other words, those in power, and wishing to gain or retain it, have always called on the creativity of the torturer to meet their ends; but along the way, they have also co-opted the work and inventiveness of many others who would have deeply regretted the contributions of their creativity. The inventor of pepper-spray is famously said to have been horrified at its use by police against protestors.[7]

11.4 Conclusion

Historically there were times, particularly in pre-enlightenment Europe, when creativity was regarded with fear (Becker 2001). Those who claimed it or engaged in it without attributing it to God were vilified or incarcerated. This all changed dramatically in the nineteenth and twentieth centuries, albeit in very specific domains such as art, literature and music or industrial invention. The rhetoric of creative genius found itself pitted against that of democratic creativity. And, in many ways, those of us who have been educators in the twentieth and twenty-first centuries have tried in the face of increasingly regimented and surveillant education systems to encourage learning through creative means rather than the stultifying absorption and regurgitation of bits of knowledge during standardized assessment tests. But we do, nevertheless, need to reflect on creativity within its many private and public, historical and contemporary contexts.

As will have become clear during my previous writings and specifically in this chapter, I do not take a singular stance towards creativity or view it as an absolute desirable or social good in every circumstance. Both as process and product, and in many of the rhetorical domains named by our 2006 monograph as well as in several

[7] https://www.theatlantic.com/technology/archive/2011/11/ive-created-a-monster-on-the-regrets-of-inventors/249044/

that we did not venture into, the contexts of deployment, strategic priorities for which creativity is engaged and goals of use of creative products and systems need to be interrogated alongside and in tandem with an understanding of creativity as inhabiting a series of rhetorics. Some creativity – such as that discussed in the opening sections of this chapter – is either easily recognised or easily erased and dismissed based on the social status and political goals of the creative individuals and groups. All of those practices and products from democratic campaigns to resource conservation and assemblage remain well within the liberal normative frame. Yet a vast swathe of creativity remains under-theorised or wrongly theorised simply as technological innovation and/or wealth production because it does not fit so easily into a liberal normative framework.

The connections between capitalism (now neoliberalism), military occupation, colonisation, carceral systems and social discrimination are fairly clear. The complex ways in which creativity has been harnessed and exploited in order to further the goals of those in power in political and economic domains are not difficult to evidence, particularly in historical texts about colonisation, the industrial revolution and the first and second world wars. However, the importance of reading creativity transversely and historically across these systems is increasing both in order to understand them and what makes them so attractive to some people but also in order to combat, take them down and replace them with alternatives.

With precisely this aim in mind, and writing about the capitalist epoch in *The Communist Manifesto,* Marx was very clear:

> The bourgeoisie cannot exist without *constantly revolutionising the instruments of production*, and thereby the *relations of production*, and with them the whole *relations of society*. Conservation of the old modes of production in unaltered form, was, on the contrary, the first condition of existence for all earlier industrial classes. Constant revolutionising of production, uninterrupted disturbance of all social conditions, everlasting uncertainty and agitation distinguish the bourgeois epoch from all earlier ones. All fixed, fast-frozen relations, with their train of ancient and venerable prejudices and opinions, are swept away, all new-formed ones become antiquated before they can ossify. All that is solid melts into air, all that is holy is profaned .... (1948, chapter 1. Emphasis added).

The problematic in this passage – of capitalism as a creative and damaging force, of creativity and destruction as a kind of dialectic – raises a number of key questions which are encompassed by but not confined to: Why do so many of us baulk at acknowledging the ways in which creativity and destruction are often yoked together in practice? So let's conclude this chapter with a series of questions that hopefully can scaffold future research into creativity that plays a less than salubrious role in contemporary societies.

Why, in an era inundated by the harvesting of big data that has been called surveillance capitalism (Zuboff 2019) and of vigilante publics (Banaji 2018) whose internalisation of the logics of violent authoritarianism lead to multiple exclusions, lynchings and perversions of justice set against the backdrop of the resurgence of fascist political elites from Turkey and India to Israel and the US (Berberoglu 2021; Glassman 2019) is it still so rare to find discussions of torture and bombing campaigns, or deep-fakes, systematic disinformation and viral propaganda as *political creativity*? And why, particularly amongst those of us who consider ourselves politically liberal or socialist is there so much denial about the aspects of rampant capitalism that many citizens find attractive – particularly the aura of self-commodification and producing the self through competition? The answers to all of these questions suggest a deep denial that creativity and devastating harm can go hand in hand, and a dangerously myopic belief in the moral and normative worth of all forms of creativity. While we debate and research further the theoretical and definitional implications of creativity that is complicit in or actively enables dispossession, dehumanisation and violence, my main aim here has been to warn that ethics and ideology cannot be separated from the teaching and learning of creative praxis without dire consequences. Some of those consequences are already becoming clear in the neglect of environmentally friendly low-tech forms of creative survival in the global south, and in the devastation wrought to our environment through the fetishization of the digital realm which centres an extractive relationship with minerals such as Coltan and the communities which must mine it and turn it into smartphones. This is, quite certainly, a discussion to be continued.

References

Banaji, S. (2008). The trouble with civic: a snapshot of young people's civic and political engagements in twenty-first-century democracies. Journal of Youth Studies, 11 (5). pp. 543-560.

Banaji, S. (2017). *Children and media in India: Narratives of class, agency and social change*. London and New York: Routledge.

Banaji, S. (2018) *Vigilante Publics*: Orientalism, Modernity and Hindutva Fascism in India, Javnost - The Public, 254: 333-350

Banaji, S., & Bhat, R. (2019). *WhatsApp vigilantes: an exploration of citizen reception and circulation of WhatsApp misinformation linked to mob violence in India*. Department of Media and Communications, London School of Economics and Political Science, London.

Banaji, S. and Buckingham D. (2013) *The Civic Web: Young People, the Internet and Civic Participation*. Massachussets MA: MIT Press.

Banaji, S., Burn, A., & Buckingham, D. (2006). *Rhetorics of creativity: A review of the literature*. Centre for the Study of Children, Youth and Media, Institute of Education, University of London.

Banaji, S., & Mejias, S. (2020). *Youth active citizenship in Europe: Ethnographies of participation*. Basingstoke: Palgrave Macmillan.

Banaji, S., Mejias, S., Kouts, R., Piedade, F., Pavlopoulos, Vassilis, Tarkova, I., Mackova, A. and Amnå, E. (2018). Citizenship's tangled web: Associations, gaps and tensions in formulations of European youth active citizenship across disciplines. *European Journal of Developmental Psychology, 15*(3), 250-269.

Becker, D. (2001). The Association of Creativity and Psychopathology: Its Cultural-Historical Origins, *Creativity Research Journal*, 13(1): 45-53.

Berberoglu, B. (2021). (Ed)*The Global Rise of Authoritarianism in the 21st Century: Crisis of Neoliberal Globalization and the Nationalist Response*. London and New York: Routledge.

Dean, J. (2013) The Limits of the Web in an Age of Communicative Capitalism, You Tube: https://www.youtube.com/watch?v=Ly_uN3zbQSU Accessed 1 May 2021.

Garrett, P. M. (2021). Getting 'creative' under capitalism: An analysis of creativity as a dominant keyword. *The Sociological Review*, *69*(1), 21–36.

Glassman R.M. (2019). The Resurgence of Fascism in the Mid-Twenty First Century. In: The Future of Democracy. Springer, Cham

Huang, Y., Felix Juefei-Xu, Qing Guo, Xiaofei Xie, Lei Ma, Weikai Miao, Yang Liu, Geguang Pu (2020). FakeRetouch: Evading DeepFakes Detection via the Guidance of Deliberate Noise. ArXiv labs. Available to download from https://arxiv.org/abs/2009.09213 Last accessed 6 May 2021.

Lerner, N (1986). UN Convention on Torture. Israel Yearbook on Human Rights, Volume: 16: 126–142.

Lowther, A., Mitchell, B., Puccio, G. & Schwagler, N. (2020). Embedding creativity in professional military education: Understanding creativity and its implementation. Available from https://thestrategybridge.org/thebridge/2020/8/10/embedding-creativity-in-professional-military-education-understanding-creativity-and-itsimplementation

Macková, A., Mejias, S. & Macek, J. (2020). When facebook is (Not) enough: Hybridity in the media and political strategies of leftist youth organisations. In: *Youth Active Citizenship in Europe: Ethnographies of Participation*. Springer International Publishing, pp. 159-187.

Marx, K. (1948). *The Communist Manifesto*. Available for download at https://www.marxists.org/archive/marx/works/download/pdf/Manifesto.pdf

Meng B. (2018). *The Politics of Chinese Media: Consensus and Contestation*. Basingstoke: Palgrave Macmillan.

Raji, Inioluwa Deborah, Andrew Smart, Rebecca N. White, Margaret Mitchell, Timnit Gebru, Ben Hutchinson, Jamila Smith-Loud, Daniel Theron, and Parker Barnes. (2020). Closing the AI Accountability Gap: Defining an End-to-End Framework for Internal Algorithmic Auditing. In Conference on Fairness, Accountability, and Transparency (FAT* '20), January 27–30, 2020, Barcelona, Spain. ACM, New York, NY, USA. https://doi.org/10.1145/3351095. 3372873

Ross, A. (2009). *Nice Work If You Can Get It: Life and Labor in Precarious Times*. New York; London: NYU Press.

Van Dyck, J. (2018). *The Platform Society: Public Values in a Connective World*. Oxford: Oxford University Press.

Zuboff, S. (2019). *The Age of Surveillance Capitalism: The Fight for a Human Future at the New Frontier of Power*. Profile Books.

Shakuntala Banaji, PhD, is Professor of Media, Culture and Social Change in the Department of Media and Communication at the London School of Economics and Political Science. She is the winner of multiple awards for teaching and has run and participated in a series of multi-country cross-national research projects on creativity, democratic participation, disinformation, new media and technologies, children and young people. She is the author or co-author of more than 75 papers, reports and chapters as well as 6 books including Social Media and Hate (Routledge 2022).

Chapter 12
Teaching (for) Experimental Creativity

David W. Galenson

Abstract Our society has long given its greatest acclaim to the sudden and dramatic breakthroughs of young conceptual innovators. This has come at the expense of undervaluing the less conspicuous, but no less important, discoveries of older experimental innovators. The internet revolution, and the associated ubiquity of personal electronic devices, have served to reinforce this disproportionate emphasis on rapid conceptual advances. I argue that our educational system needs to teach students both greater appreciation of experimental creativity and the skills necessary to practice it. Our students, and our society at large, should be made to recognize that the greatest accomplishments in all intellectual domains have been, and should continue to be, the products not only of sudden flashes of genius, but also of long and painstaking study and research.

One of the most widespread beliefs about creativity is that it is primarily, or even exclusively, associated with youth. The stereotype of the brash and iconoclastic young genius as the dominant source of important innovations has long had a firm hold on the popular imagination. Nor is this belief restricted to the general public, for it is also shared by many scholars. The psychologist Harry Lehman (2017, pp. 330–331) concluded that "superior creativity rises relatively rapidly to a maximum which occurs usually in the thirties and then falls off slowly." Old age was anathema to creativity, for "when a situation requires a new way of looking at things, the acquisition of new techniques or even new vocabularies, the old seem stereotyped and rigid." Other psychologists have agreed, as more recently Colin Martindale (1989, p. 221) observed that "in general, a person's most creative work is done at a fairly early age," and Dean Simonton (1990, p. 326) opined that "creativity seems to peak in early to middle adulthood." Scholars in other disciplines

D. W. Galenson (✉)
University of Chicago, Chicago, IL, USA

Universidad de CEMA, Buenos Aires, Argentina
e-mail: galenson@uchicago.edu

D. Henriksen, P. Mishra (eds.), *Creative Provocations: Speculations on the Future of Creativity, Technology & Learning*, Creativity Theory and Action in Education 7, https://doi.org/10.1007/978-3-031-14549-0_12

have also expressed similar beliefs. Thus, the economist Paul Romer declared that "young people tend to be more innovative, more willing to take risks, more willing to do things differently," (quoted in Taylor, 2010) and the biological scientist Francis Collins reflected that "the researchers in the early stages of their careers tend to be the ones with the fire in the belly. They are not afraid of if tackling the really hard problems" (quoted in Lehrer 2010).

This belief in the connection between youth and creativity has if anything been strengthened in recent decades, as the myriad technological discoveries that have contributed to the rise of the internet and the ubiquity of consumer electronics have created a prominent new class of brash and youthful billionaires. Silicon Valley worships youth. For example, Mark Zuckerberg declared that "young people are just smarter," (quoted in Samuel, 2017) and his Palo Alto neighbor Vinod Khosla observed that "people under 35 are the people who make change happen. People over 45 basically die in terms of new ideas" (Wadhwa 2011). As a result, the belief in the dominance of the young in technological change may be at its highest point.

A basic problem for our society, from which a series of other problems follow, is that the belief that creativity requires youth is false. This belief is a classic case of mistaking part of a phenomenon for the whole. The bold leaps of fearless young *conceptual* innovators are obviously an important form of creativity. But they are *not* the *only* form. For there is another, very different type of creativity, in which important novel discoveries emerge gradually and incrementally from extended cautious explorations by older *experimental* innovators.

Conceptual and experimental innovators differ in both the nature of their creativity and the means by which they achieve it. Conceptual innovations express novel ideas. Ideas can arrive fully formed, so there is generally a clarity to conceptual innovation: a new work can fully express the new idea.

In contrast, experimental innovators want to express perceptions. They work inductively to do this, by developing new techniques based on extended observation and experimentation. Their goals are imprecise, and their uncertainty leads them to proceed incrementally, by trial and error. Experimental innovations are rarely embodied in a single work, but usually appear piecemeal in a series of related works.

The differing creative life cycles of these innovators are a consequence of these differences in method. The most radical conceptual innovations are typically made by the young—recent entrants into a discipline, who are free to make discrete departures because they have not had time to become accustomed to following the domain's established conventions. In contrast, the late innovations of experimentalists are generally their most powerful, for they are based on greater accumulated knowledge and technical mastery developed over longer periods.

Yet if experimental innovation is as important as conceptual, why—as claimed above—is it so often overlooked? The answer lies in the differing nature of the two types of creativity. Conceptual innovations are often bold and conspicuous. For example, the poet William Carlos Williams lamented that T.S. Eliot's *Waste Land*, "wiped out our world as if an atom bomb had been dropped upon it" (Williams 1951, p. 174). Other atomic bombs dropped by young conceptual innovators include Picasso's *Demoiselles d'Avignon*, Stravinsky's *Rite of Spring*, Joyce's *Ulysses*,

Welles' *Citizen Kane*, Fitzgerald's *Great Gatsby*, Godard's *Breathless*, Plath's *Ariel*, and Dylan's "Like a Rolling Stone." All of these conceptual innovations differed radically from anything that had preceded them, and caused sudden changes in the way many practitioners in their disciplines worked.

In contrast, even the greatest experimental innovations often arrive inconspicuously and almost imperceptibly. There is no individual work that announced the innovations of Charles Dickens, Paul Cézanne, Virginia Woolf, Robert Frost, John Ford, Elizabeth Bishop, Philip Roth, or Saul Bellow. No single painting embodied the visual innovations of Cézanne, but instead a series of paintings that span the final decades of his life, each of which differs only marginally from the one that preceded it. One result of this was that the impact of Cézanne's discoveries on younger artists was delayed, until they could see a number of his canvases together.

In some cases, experimental creativity has been overlooked for an extended period because of its very success. One goal of many great experimentalists has precisely been to make their innovations unobtrusive. For example, Alfred Hitchcock, now widely regarded as the most important American film director, was for much of his career considered merely a commercial hack by critics who did not understand the subtlety of his technical innovations. And their unobtrusiveness was deliberate, for Hitchcock maintained that "Technique that calls itself to the audience's attention is poor technique. The mark of good technique is that it is unnoticed" (Gottlieb 1995, p. 208). Robert Frost similarly suffered for much of his career from the charge that his poetry was not artistic, but that it was simply a reproduction of everyday speech, because of his very success in on mastering "an art that conceals art" (Dickstein 2010, pp. 23–24). Irving Berlin was often dismissed by critics as an unsophisticated songwriter, whose work lacked depth and maturity. Yet his songs were recorded by such jazz greats as Benny Goodman, Duke Ellington, and Earl Hines, as well as the singers Willie Nelson and Pete Seeger. The skill of these and other experimental innovators was long overlooked ironically because of their very success in pursuing the experimental values of clarity and simplicity: to capture reality directly.

Education should play a key role both in increasing public appreciation of experimental creativity and in preparing students to become experimental innovators. In general, today's education is seriously deficient in accomplishing these goals. Interestingly, some understanding of this can gained by considering education in an earlier time, and the case of a great experimental innovator.

As a child, Charles Darwin loved the outdoors, and hated school: the local school he attended from the age of nine "as a means of education to me was simply a blank." The school's curriculum emphasized the classics, taught by rote, and Darwin had no interest: "When I left school, I was for my age neither high nor low in it; and I believe that I was considered by all my masters and by my Father as a very ordinary boy, rather below the common standard in intellect." An indication of Charles' true intellectual interests came when he and his older brother Erasmus equipped what they called their Laboratory in a spare closet, where they did chemical analysis of minerals (all quotes in this paragraph from in Browne 1995, pp. 12–29).

Darwin's father, a prosperous physician, sent Charles to Edinburgh University to become a doctor, but Charles had no interest in his courses, which were again taught largely by rote, and left without a degree. Now his father was angry, telling Charles "You care for nothing but shooting, dogs, and rat–catching, and you will be a disgrace to yourself and all your family." Charles could hardly object: "He was very properly vehement against my turning an idle sporting man, which then seemed my probable destination" (all quotes in this paragraph from in Browne 1995, pp. 65–89).

His father then sent Charles to Cambridge, to prepare for a career as a country clergyman. Yet as a biographer observed, "the most striking aspect of his career at Cambridge University was the total lack of any engagement with academic work." But a friend introduced Charles to a hobby—hunting for beetles in the local fens—that became his passion. And it was through his zeal in the pursuit of rare types of beetle that Charles met the professor of science who became his mentor, and later recommended him for a position on a surveying voyage to South America (quotes in this paragraph from in Browne 1995, pp. 96–115).

Darwin described John Henslow as "quite the most perfect man I ever met." Initially a professor of botany and later of mineralogy, Henslow was committed to studying nature in the field. His popular lectures were supplemented with practical instruction in recognizing and dissecting plants, and with field excursions to observe nature firsthand. Darwin was excited to use a microscope to look at the anatomy of plants, and to discover their characteristics himself rather than memorizing them from books. Here at last was an experimental teacher who could guide and challenge an experimental student, recognize his enormous talent, and start him on his career as a scientist (quotes in this paragraph from in Browne 1995, pp. 124–40).

It was on the *Beagle* that Darwin received "the first real training or education of my mind. I was led to attend closely to several branches of natural history, and thus my powers of observation were improved" (Darwin 2005, p. 64). And it was on the *Beagle* that it became clear that Darwin's earlier lack of academic success was not due to any innate lack of industriousness or intellectual curiosity, as during the eventual five years of the voyage he wrote 770 pages of a diary, 1383 notebook pages on geology, 368 pages on zoology, collected 1529 samples of species in spirits, and 3907 labelled animal skins and bones (Desmond and Moore 1992, p. 189).

Darwin was 27 when the *Beagle* returned to England. He would famously work for another 23 years before he published *On the Origin of Species*. During that interval, the four volumes he published on barnacles, and the thousands of letters he wrote and received in the course of his research on evolution, were just parts of the vast accumulation of evidence for one of the most consequential experimental research projects ever undertaken.

Late in his life, Darwin gave a concise statement of his formula for creativity, in a letter congratulating his youngest son on passing a college exam. The boy was not a distinguished student, and Darwin could clearly identify with him. His encouragement to his son stressed that creativity did not depend solely on intelligence:

> I have been speculating last night what makes a man a discoverer of undiscovered things, and a most perplexing problem it is. Many men who are very clever—much cleverer than discoverers—never originate anything. As far as I can conjecture the art consists in habitually searching for causes or meaning of everything that occurs. This implies sharp observation and requires as much knowledge as possible of the subject investigated (Darwin 2005, p. 132).

Charles Darwin's educational career in the early nineteenth century is so distant in time and content that it might be dismissed as irrelevant to any consideration of the problems with education two centuries later. In fact, however, Darwin's experience is a powerful example of how an educational system failed to develop the intellectual potential of a student who needed training suitable for an inductive, experimental innovator. Thus Shrewsbury School, Edinburgh University, and Cambridge University all failed to provide a formal curriculum that taught the skills—sharp observation and as much knowledge as possible of the natural world—that would eventually make Darwin the greatest scientist of the nineteenth century. Fortunately, Darwin's curiosity about the natural world was so great that he effectively taught himself entomology in the marshes around Cambridge, and his talent for this was such that he attracted the attention, and soon the friendship, of a brilliant young professor of science who became a key figure in guiding his true education.

Education today of course bears little resemblance to that of Darwin's youth: few students today, for example, are required to study classics. Yet in a crucial respect, which was perhaps the greatest problem for the young Darwin, I believe that too much of education today is similar. This is in teaching mechanically, by rote. In what follows I will discuss only higher education, and primarily one discipline. Yet I believe that the attitudes I will describe are widespread in our society, and that as a result the problems I see are general to other levels of education, and to other disciplines.

I was an undergraduate economics major in the early 1970s, a graduate student in economics in the late '70s, and I have taught in economics departments since 1978. I have seen a consistent trend in the way academic economics is taught and practiced throughout my 50 years in the discipline, and I believe that the basic trend is a continuation of changes that had begun several decades prior to my own experience. I might note that my discussion of the trends in economics education is not based on systematic evidence, but I do not believe it is controversial, and I think it could be documented without great difficulty.

Most simply, at both undergraduate and graduate levels, economics today is taught highly mechanically. There has been a longstanding trend toward greater use of mathematics: this is generally described, by both teachers and textbook salesmen, as an increase in rigor. The rote element is a consequence of the pervasive use of mathematics. Classroom lectures in many fields of economics consist largely of mathematical derivations and proofs. And even more mechanically, the homework assignments for most economics courses consist of mathematical problem sets.

Neither the classroom derivations nor the myriad problem sets are of interest or significant intellectual value to experimentally inclined students. In effect, the constant emphasis is on abstract propositions and techniques. Theory and econometrics are of course the basis of most good applied economics—analysis of problems in the outside world. But few courses actually demonstrate how these tools are applied in practice. The frequent result is students who have been taught theoretical tools, but have little or no idea how those tools can be used for any practical purpose. To these students, economics is little more than a field of mathematics and statistics.

Students have had little say in this trend: those unable or unwilling to do the necessary mathematics have chosen other majors. Yet I believe there is a recent technological factor that underlies the lack of student opposition to the absence of applications of the mathematical tools they are required to learn. This is the growth of the internet, and the associated availability of inexpensive personal computers. Recent generations of students have had a growing belief that any necessary knowledge can be acquired by googling. The idea of studying written texts to gain deep-knowledge of a subject, or evidence relevant to one, is increasingly seen as antiquated. Indeed, the words *knowledge* and *evidence* have largely been replaced in students' vocabularies by the term *information.* Students today often believe that college is a place to learn technologies, that can be applied later in their careers simply by adding instant information. Unfortunately, this belief ignores the fact that application of these tools to real problems is itself a sophisticated process that requires extended training and practice, and that without this training the tools are of little practical value. Many students are thus not only not trained to practice experimental creativity, but are not educated to understand and appreciate what it is and why it is valuable.

Economics students who need to learn economics inductively, with applications, have to seek out teachers who will provide this. How difficult it is to find these teachers depends on the department in question. Nor is this problem unique to economics. Scholars in other disciplines have expressed similar concerns. So for example the physicist Matthew Fleenor observed that "While physics is often viewed as a discipline overly populated with conceptual innovation, undergraduate instruction within the discipline is historically associated with qualities that hinder creativity, which may be an especially harsh environment for experimental innovators" (Fleenor 2018, p. 73). The chemist John Moore was equally concerned about the plight of experimental students:

> It seems to me that this division of creativity into two classes also applies to the students we work with on a daily basis. Some are quick studies and apparently assimilate new ideas effortlessly. Others have to work really hard to develop thorough understanding. It is easier to deal with the former. They are the students of whom I have heard it said that they would learn whether they were taught well or badly, so why worry so much about teaching? Far more rewarding, though, are the students who work tenaciously to learn. They (and we as well) take great satisfaction from their hard–earned knowledge. It is encouraging to have support for the idea that such students may be just as creative as their more immediately brilliant contemporaries. Even if it requires many years for their creativity to reach fruition, they may well contribute as much or more to society. There is a strong tendency to think that only the quick studies are really creative, but we ought not sell the experimental innovators short during our brief time as their mentors. To do so might lose much talent that could otherwise improve our society (Moore 2007, p. 7).

Albert Einstein's explanation for his success was simple: "Imagination is more important than knowledge" (quoted in Viereck 1930, p. 447). Charles Darwin's was very different: "My industry has been nearly as great as it could have been in the observation and collection of facts" (Darwin 2005, pp. 114–15). Einstein was an archetypal conceptual innovator, Darwin an archetypal experimental innovator. Both of these types of innovators have made enormous contributions to the society and economy of the modern era. But as discussed in this paper, recently there has been a disproportionate tendency to celebrate one of these types at the expense of the other. And our educational systems have tended to incorporate the same bias, favoring conceptually oriented students, and stressing the importance of conceptual contributions.

How to eliminate these biases is a large and complex problem, which involves deeply held attitudes and longstanding practices. Nor is there likely to be any simple or rapid solution: witness the long and arduous campaigns against racism and sexism that have been waged for decades and remain far from complete. In large part, the elimination of the prejudice against experimental creativity will parallel those other campaigns, for it will require a direct confrontation of our society's ageism.

But the first step in solving a problem is recognizing it. This paper has called attention to the need for our society to appreciate the importance of experimental creativity, and to understand that our educational system has a critical role in teaching students to engage in it. And an important part of the campaign to recognize and foster experimental creativity will be to perceive how current trends in technology are exacerbating these problems. Thus, the way we use the internet and personal computers is currently tending to constrain individual creative development, because so much of social media is aimed at instant gratification, and encouraging short attention spans.

Technology is having much the same effect in educational settings, because even the texts students read online are often peppered with deliberately tempting and distracting hyperlinks. The immediacy of the myriad opportunities of the internet at every moment runs counter to the extended concentration needed to master a serious text, whether article or book. In general, electronic tools are very often used in ways that run counter to the skills Darwin cited as critical to experimental creativity, as the vast store of texts and images in cyberspace is taken by students to make unnecessary their own cultivation of sharp observation and patient accumulation of knowledge of a domain. Experimental creativity is not a lost cause, but it is not thriving today, and it is important that our society takes steps to strengthen it, to counter the powerful forces of current trends in technological change.

References

Browne, J. (1995). *Charles Darwin: Voyaging*. Princeton University Press.
Darwin, C. (2005). *The autobiography of Charles Darwin*. W.W. Norton.
Desmond, A. & Moore, J. (1992). *Darwin*. Warner Books.

Dickstein, M. (2010). *Robert Frost.* Salem Press.
Fleenor, M. C. (2018). Cultivating experimental innovation within undergraduate physics majors. *Global Education Review, 5*(1), 73–87.
Gottlieb, S. (1995). *Hitchcock on Hitchcock.* University of California Press.
Lehman, H. C. (2017). *Age and achievement.* Princeton University Press.
Lehrer, J. (2010, Feb, 19). Fleeting youth, fading creativity. *Wall Street Journal.* https://www.wsj.com/articles/SB10001424052748703444804575071573334216604
Martindale, C. (1989). Personality, situation, and creativity. In J. A. Glover, R. R. Ronning. & C. R. Reynolds (Eds.), *Handbook of creativity* (pp. 211–232). Springer.
Moore, J. W. (2007). The art and science of science and art. *Journal of Chemical Education, 84*(1), 7.
Samuel, L. (2017, Oct. 2). Young people are just smarter. *Psychology Today.*
Simonton, D. K. (1990). Creativity and wisdom in aging. In J. E. Birren & K. W. Schaie (Eds.), *Handbook of the psychology of aging, (3rd ed)* (pp. 320–329). Academic Press.
Taylor, T. (2010, Feb, 26). Bright Old Things. *Globe and Mail.* https://www.theglobeandmail.com/report-on-business/rob-magazine/bid-idea-bright-old-things/article4311908/
Viereck, G. (1930). *Glimpses of the great.* Macauley.
Wadhwa, V. (Dec. 2, 2011). The case for old entrepreneurs. *Washington Post.* https://wadhwa.com/articles-list/2011/12/2/washington-post-the-case-for-old-entrepreneurs
Williams, C. W. (1951). *The autobiography of William Carlos Williams.* Random House.

David W. Galenson is Professor of Economics at the University of Chicago; Academic Director of the Center for Creativity Economics at the Universidad del CEMA, Buenos Aires; and a Research Associate of the National Bureau of Economic Research. He is also Contributing Editor to The Art Economist magazine and a blogger for the Huffington Post. He has been a visiting professor at the California Institute of Technology, Massachusetts Institute of Technology, the University of Texas at Austin, the École des Hautes Études en Sciences Sociales in Paris, and the American University of Paris.

Chapter 13
How Not to Kill Creativity?

Yong Zhao

Abstract When we are seeking approaches to foster creativity, we should also consider removing mechanisms that can discourage students to become more creative. If we accept the idea that all humans are born creative but they become less creative as they grow up because of their experiences, we can find that schools are a major mechanism that reduces creativity. All children attend schools but schools, as they have existed, are built to impose on all students a homogenous curriculum implemented through homogenous pedagogy. The homogeneity of school experiences force students to lose their interest and confidence in becoming creative as they are rewarded for being compliant instead of creative. This chapter discusses how schools can kill creativity. It also discusses how technology can both support and hinder the development of creativity in schools.

Keywords Creativity development · Schooling · Technology · Innovation

People in education have a very bad habit. Whenever something is said to be important, we try to teach it. Social and emotional learning (SEL) has gained importance so there are programs to teach it. Computational thinking is now considered important, so schools should teach. Growth mindset is important, bring it to the classroom. So are grit, resilience, and a host of other traits and abilities. If they are important, we must teach them.

When something has garnered enough importance in the minds of policy makers, it becomes something to teach. Before it can be taught in every school, a system has to accept it. Once accepted, it gets into the national or state curriculum framework or standards. Once it is in the standards, assessment comes along. Once assessment is made, schools begin to teach it and apply assessment to students. With assessment, schools can feel good about having incorporated the "it" into the curriculum and assessed this important thing.

Y. Zhao (✉)
School of Education, University of Kansas, Lawrence, KS, USA

D. Henriksen, P. Mishra (eds.), *Creative Provocations: Speculations on the Future of Creativity, Technology & Learning*, Creativity Theory and Action in Education 7,
https://doi.org/10.1007/978-3-031-14549-0_13

But teaching may be exactly what is not needed. In fact, teaching may just be the opposite of what is needed. For example, SEL is ultimately concerned with student psychological and social wellbeing. But standardized schooling, testing, over teaching, and centralized homogenization destroy student well-being, deprive them of the opportunities to play and interact with their peers, and place them under great pressure. The irrelevant lectures, meaningless homework, and nonsensical assessments gave students no time to exercise self-determination, to practice self-regulation, and to develop self-awareness (Wehmeyer and Zhao 2020; Zhao 2020a). As a result, while this kind of teaching may improve test scores, it hurts social and emotional development (Emler et al. 2019; Zhao 2018b).

To improve SEL does not require teaching SEL. Adding SEL to state standards and require all students to have the same SEL skills can only make the situation worse because it places more pressure on teachers and students (Zhao 2020a). What needs to be done is to examine the situation students are in and make decisions to remove the sources of pressure and anxiety and to increase time and space for self-determination and autonomous actions.

Creativity is one of the qualities that are emerging to be important in education. It has been listed as one of the four Cs (creativity, critical thinking, collaboration, and communication) necessary for the Twenty-First Century (Partnership for 21st Century Skills 2007). It is listed in different curriculum frameworks (e.g., (Australian Curriculum Assessment and Reporting Authority 2010)). It is part of the Common Core curriculum in the United States (Common Core State Standards Initiative 2011). Writings about creativity in the education literature have also been on the rise.

This is great news for creativity but at the same time it is worrisome. Creativity has not yet been made a required subject or course in schools, but for how long can we resist it? What if it is made a required subject? What if schools are forced to teach and assess creativity in the future? What happens if creativity is taught in schools as a required subject?

13.1 Dangers of Teaching Creativity

Teaching creativity as a school subject like math and science can be very dangerous. It can work to hurt the development of creativity in the population and in individual students. There are a number of ways this can happen.

First, when creativity is taught as a required course in schools, there must be a standard, a framework, or a sequence and scope. Otherwise, what is there to teach? Teachers must follow the standard, the framework, and the sequence and scope to plan lessons, to implement the lessons, and to make sure that students are indeed taught creativity. In a sense, standardization is required for classroom instruction of creativity following however creativity is thought and defined. Such standardization actually kills creativity effectively.

Creativity comes from the human desire to invent something new. It comes from the human nature to combine what is available and develop new possibilities. No matter how it is defined, creativity is individual. It is naturally born and can be suppressed or supported with experiences. Human beings, because of their interest, confidence, value, and abilities, can be more or less creative in different domains and contexts. They can also be more or less creative with different content and products. Some people may be more creative in arts, others in music, still others in math. Some people may be more interested in creating new sculptures, others may be interested in writing poetry, and still others may want to explore new possibilities in science.

When they are taught to be creative, they are in essence forced to exercise their creativity for the purpose of getting a grade or going through a course. They are not necessarily exercising their creative potentials following their own desire and interest. Moreover, a course in school, if it is like most courses in schools, is about having students go through a standardized process led by the teacher. Thus a creativity course in schools would necessarily focus only on a narrow set of domains of interests. It is impossible for capture all possible areas that students can be interested in. Thus, for many students in the class, the only motivation they may have is to pass this course.

As a result, those students who are interested in pretending to have creativity following the course requirement may show some signs of furthering their creativity. But those who are not interested in doing so or are unable to do so will likely see their creativity decline. They would also be the ones who are considered less creative or lacking the ability to become more creative. Such decisions can be detrimental for students interested in and capable of being creative in the domains not included in the course. This is very much like students who are talented in non-school subject areas have been cast away as losers or at-risk (Zhao 2018a). These students could lose interest in schools and also teach themselves as helpless.

Second, teaching creativity as a course in schools likely requires the teacher to teach certain strategies and actions that have been ostensibly proven to be effective in promoting creativity in the literature. While there are certainly strategies and tactics that have been identified as helping creativity growth in some areas, it is questionable if they apply to all areas. There is much debate about whether creativity is generic or domain specific. Without going into the details, I am pretty sure that there is no one set of strategies and actions that work for all students in all domains across all ages (Zhao 2018b).

If there are no strategies and tactics that promote creativity for all students in all domains, whatever the course uses can misguide some students or waste their time because the course cannot possibly cover all areas. For example, a student interested in drama may have little to learn from strategies that promote creativity in science or engineering. Or a student who is excellent and interested in writing novels can be misguided by strategies that may be helpful for students interested in being creative in math. Similarly, a student who is already quite creative in her own ways can be misled by the course content because she does not need it anymore. For a similar reason, a student who is way behind may not even understand what is being taught in the course.

Third, teaching creativity in a course can also hurt student development in creativity by taking time away from experiences that can actually help. Creativity is complex. One does not only need to be creative, to come up with novel ideas, he needs knowledge and skills to be meaningfully creative. Considering the four levels of creativity (Kaufman and Beghetto 2009), we can say that everybody can have mini C but to become professionally creative, she needs professional knowledge and to understand the professional landscape. Even little C requires knowledge and skills. Thus teaching creativity in isolation from knowledge and skills is mostly a waste of time. It may be an enjoyable exercise for some people but it can hardly make a person more creative.

Moreover, creativity is needed all the time. It is not a one-time deal. Thus a course taught at first grade to boost creativity may have lost its impact when the child enters third grade. So far we do not know how long the impact of creativity teaching lasts. We also do not know if creativity dissipates with time. Thus we need to ask the question is we need to teach creativity every semester or every year or every 5 years. Or perhaps, once you teach it, creativity lasts forever.

13.2 Dangers of Large-Scale Assessments

Large-scale assessments, assessments that are applied to a large population as in the world, a nation, or state educational system, have been popular for the past few decades. Internationally we have the Programme for International Student Assessment (PISA) and Trends in International Mathematics and Science Study (TIMSS) besides others. Many countries and states have their own standardized large-scale assessments for accountability or selection of students. Typically, these assessments can be high stakes, having significant consequences on students or educators. Even PISA and TIMSS, technically having no consequence on anyone, actually do have significant impact on policy makers in different countries.

Large-scale assessments can have many problems (Emler et al. 2019). One of the problems is that they can hurt the development of creativity. So far there is no large-scale assessment of creativity and almost all of them are assessments of abilities in different academic subjects such as math, reading, and science. These assessments actually squeeze creativity out of schools (Zhao and Gearin 2016).

The large-scale assessments typically result in a process of standardization and centralization in schools. They start with frameworks of curriculum or assessment, which dictates what is to be assessed and taught in schools. The past few decades saw a significant increase in the centralization of curriculum standards and frameworks. Australia and the United States, for example, both countries that had historically made education a state responsibility, came up with national standards (Australian Curriculum Assessment and Reporting Authority 2010; Council of Chief State School Officers and National Governors Association Center for Best Practices (NGA Center) 2010).

Centralization and standardization make a powerful process to exclude creative people instead of helping them to be more creative (Zhao and Gearin 2016). Large-scale assessments, particularly when they are high stakes, require all students to do well. But human beings, as natural born and diverse learners due to interactions between nature and nurture (Lewontin 2001; Ridley 2003), cannot possible be the same. Thus as schools and teachers work very hard to make sure their students do well on tests, they limit opportunities for students to pursue their creativity, interests, and talents (Bonawitza et al. 2011; Nichols and Berliner 2007; Zhao 2018b).

Teaching to the tests has been a common practice in schools, which resulted in only subjects and content included in assessments to be included in the curriculum that students can experience. Over the past few decades, many schools have shifted attention to math and reading, reducing time for other subjects, even recess (Emler et al. 2019). Consequently students who are not interested in reading and math or doing well on reading and math assessment can be deprived of the opportunity to develop their own talents and interests. They are deprived of the chance to trigger their natural born talents, to meet their needs, to excite their desires. Their creativity has no place to be cultivated and exercised (Tienken and Zhao 2013).

13.3 Dangers of the PISA Creativity Assessment

Recently, PISA has decided to add creativity to its assessment package (OECD 2019). The assessment was to be delivered in 2021 but due to the COVID-19 pandemic, the assessment was postponed to 2022. The assessment is intended to be taken by 15 years old students in different education systems in the world and has attractive goals:

The PISA 2021 creative thinking assessment will provide policymakers with valid, reliable and actionable measurement tools that will help them to make evidence-based decisions. The results will also encourage a wider societal debate on both the importance and methods of supporting this crucial competence through education. This work in PISA is connected to another OECD project that aims at supporting new pedagogies that can foster creative thinking. (OECD 2019, p. 5).

There is no large-scale assessment that can validly and reliably assess what it claims to assess. Such is the case with the SAT and ACT, which could only predict college success at a fairly low level. And there is no creativity assessment that can predict reasonably accurately a person's creativity in life. So what PISA has is an untested assessment for creative thinking. It plans to assess students in as many cultures and contexts as possible. The results will be numbers that PISA extracts and these numbers will be used to judge which education system is good at cultivating creativity.

That decision is scary, scarier than the usual PISA results, which compare the quality of education of different systems based on test scores in math, reading, and science. The usual results have already created an illusion of educational excellence

(Zhao 2020b). While PISA claims that it measures what matters in the Twenty-First Century, there is no evidence it does so, nor are there reasonable arguments that it does.

The potential decision that PISA will make based on the creativity assessment is likely to create another illusion of excellence in cultivating creativity. Because of the superpower of PISA and its owner the Organisation for Economic Co-operation and Development (OECD), it is likely that the PISA creativity assessment will be taken seriously in the world. It is quite possible that the creativity assessment becomes the *de facto* gold standard of creativity development in different education systems.

But the PISA assessment is not and cannot be the gold standard. It is just a test. Despite the theoretical arguments and research evidence documented in the framework (OECD 2019), the assessment is a standardized fake test of creativity. It measures how well students perform on the test items instead of authentic creativity. At best, it can tell variations of student "creativity" on the items included in the assessment, which is drastically different from the real creative potentials, creative confidence, and creative interests a student may have.

While the assessment may or may not measure creativity, it can be believed to measure creativity among education policy makers, educators, parents, and students. If they believe it measures creativity, performance on the assessment becomes important for all involved. The result could be very simple teaching to the test, which happens a lot when a test becomes important. In other words, the content of the assessment can become the core of what is taught in schools. If the assessment truly measures creativity, it may not be a bad thing to teach it. However, the assessment cannot possibly measure creativity of students in all education systems around the world.

The PISA creativity assessment could also have a negative impact on creativity development in schools around the world because of its power. It could affect how people view creativity and creativity assessment. If parents, the public, and policy makers treat the assessment seriously, we could expect educational policies and practices that push PISA-defined creativity on students and in schools, which again would be a mistake because we do not know if it actually assesses creativity.

13.4 Opportunities and Dangers of Technology

Today, technology is an unavoidable topic when talking about creativity. First, digital technology is ubiquitous for students today. Today students spend more time on digital media than in schools (Rideout and Robb 2019). Second, digital technology has become even more ubiquitous for students because of COVID-19. Every student in the United States, for example, should have access to the Internet and smart devices or computers. Third, digital technology has become an ordinary and common companion for students and teachers as a tool of learning and teaching. It is thus inevitable to think how technology may affect the development of creativity.

Technology is a double-edged sword for creativity. On the one hand, technology can drastically help the development of creativity in schools. For instance, with digital technology today, which would have cost millions 20 or 30 years ago, students can manipulate videos, audios, images, and texts easily on a smart phone or pad. They can also use resources created by others in creating their products. Moreover, there are instructional materials online that teach the skills and knowledge needed to create videos or music or other combinations of images, texts, and audios. Not only can students create meaningful products using their own materials and resources online, they can also publish them and make them available globally for anyone to use. This means that almost every student in developed countries has access to a great device that enables them to create, recreate, and publish their creative outcomes.

On the other hand, however, technology can hinder the development of creativity. First, technology can be used to manage and control students. With the power of technology today, it is easy to have all student data into a system. These data can be used to guide and plan students' educational activities and pathways. For example, standardized test scores can be used to assign students into remedial programs or retain students for a grade. But standardized tests do not test students' creativity or individual unique abilities. Assigning students to remediation programs in literacy and numeracy can be discouraging for students to develop creativity and unique talents. It also takes valuable time away from activities that can promote creativity.

Second, technology can be used to homogenize students. The current movement to use big data and learning analytics to develop personalized learning programs for students can actually homogenize students. Although the idea behind the computerized personalized learning programs seems to allow students to learn on their own, the outcomes are actual homogenization of learning because the content and standards are the same for all students. It forces students to learn the same thing at a different speed.

Third, technology can mislead educational activities to promote the development of creativity. Because of the prominence and power of digital technology in schools, educators can easily mistake it as the only technology that can be used to support the development of creativity. As a result, students' creative development can be limited to activities that are primarily based on digital technology. Students may be able to develop creativity in videos, audios, and apps, but they may not be supported when they have interest or ability in other areas such as dancing, music, or physical environments.

What schools need to do is to consider the diversity of students and the massive number of domains that need human creativity. Schools should not use technology excessively to control students' creativity or use technology to homogenize them. Instead, schools should develop activities that support individual diversity and creativity in as many domains as possible.

13.5 What We Need to Do?

Creativity is certainly important and essential for human beings to thrive in the new world where technology is poised to displace human workers from many jobs (Zhao 2012). There is no question that those who can create value, services, and products for others and the world will be more valued (Zhao 2018a). Education must find ways to help all students to become more valuably creative. However, what needs to be done is not to teach creativity courses or having creativity assessment.

What we need to do is to create an educational environment in which students can be productively creative. Less is more should be the philosophy of the new environment. What students need is not more teaching, more assessment. Instead they need time and space of their own, with proper and meaningful guidance from teachers and parents.

It is seductive to teach and assess creativity, but what is needed to cultivate creativity is space and time. Schools today are already overcrowded. A student's time is fully with prescribed curriculum, pre-planned instruction, and predetermined organization. A student has very little time and space in school to deviate from the curriculum, the social and organizational norms, and the organized classes. So we do not need to add more courses. Rather, we should be taking things out of the student experience.

The first thing to take out is large-scale assessments. Schools have been preoccupied with preparing students for these assessments. They are also very sensitive and pay a lot of attention to testing results. In some schools, much of students' experience is occupied by these assessments. COVID-19 in many ways has made many governments pause these assessments. It would be great if these assessments never return to schools.

Second, schools should cut back required courses. Almost all schools are subject to national or state curriculum, which prescribes all the school time to the teaching of required courses. Even private schools typically prescribe courses for time students have in schools. Students are asked to take the same courses based on their age. There may be elective courses, but typically they are offered by school teachers. To occupy all student time with required courses does not give students the time and space to develop their creativity in the areas they are interested in. Thus, schools need to reduce required courses so as to enable students to take control of their own learning (Wehmeyer and Zhao 2020; Zhao 2012, 2018a). Students should be allowed to take courses and engage in other educational experiences on their own.

Third, retreat from over governance and over management. Students are over governed in schools and they are also over managed. Students are the agents of learning and thus should be the owners of their learning. But their agency is taken away so students are the least powerful people in schools. To perform the job of teaching, schools also manage their students carefully. Students are always with an adult teacher, who takes care of students' attendance and monitor disciplinary issues.

Students are not given the right to decide what they want to learn, when to learn, and where to learn. Students' behaviors are also monitored all the time. In other words, they are expected to be students instead of themselves. To change the school into an educational environment that fosters creativity, schools need to become personalizable and give students the right to exercise self-determination (Wehmeyer and Zhao 2020; Zhao 2012, 2018a).

Of course, to create an educational environment that cultivates creativity does not end with simply taking things out and giving students the space they need. It also requires a number of things that schools have not been doing. First, we need to create a culture that celebrates creativity and creative individuals. To celebrate creativity and creative individuals means to encourage everyone, student and staff, to be creative in their own way. It is not to predefine a narrow set of creative behaviors but to broaden the success of schools to include success in creativity. It is to enable individuals to pursue their interests and creative capacity. It is to have places and events to showcase the creative works of students and staff. It is also for staff to support and motivate students to become creative.

The second thing we need to add is to make sure that every student has a creative strengths profile (Zhao 2018a). Education should be strength-based. Instead of focusing on fixing students' deficits, it should be on developing individual differences (Zhao 2016). In developing student' strength, it is important that every student has a profile constructed with teachers and parents to record their strength, activities to strengthen the strength, and products to demonstrate the growth of the strengths. Strength-based education is about product-oriented learning, which means all learning should start by identifying problems worth solving and end with a solution (Zhao 2012). This process requires the development and application of creativity. Thus a creative strength profile helps students to become aware of their strengths and creativity and accountable for their development.

The third thing we could add is to infuse creativity in all courses. Schools should focus on creativity as a core competence. It does not belong to one course but all courses. Students should have creativity-enhancing educational experiences all the time. All courses should be focusing on enhancing and developing student creativity, from kindergarten to 12th grade.

Finally, add online and global education to expand the opportunities for all students. Technology has become so mature that students do not need to take only courses from their school. In fact, COVID-19 has drastically changed education in schools. The short period of online or remote learning suggests that online and global learning is possible. Thus in order to meet the personal needs of each and every student, we must go beyond our own staff and courses. We need to engage students in a global learning ecosystem so that they are interacting with, learning with, and learning from anyone on the globe. They are also engaged in creative activities to create value for anyone on the globe.

13.6 Summary

Creativity is individual and diverse. Creativity is natural born but needs continuous nurturing and cultivation. For creativity to be expressed and developed, human beings need an open environment and strong encouragement. But students may have different domains where they want to be creative. Thus creativity cannot be taught as a required, standardized course for all students.

For creativity to become more than novel and random thoughts, for it to be valuable, human beings need to develop knowledge and skills. This makes creativity domain specific. But because human beings can choose which areas they may be interested and good at, creativity needs to be fostered in ways that support personalization of learning so that each and every student is able to enhance their creativity in their own unique ways.

Creativity is a lifelong quality of human beings. It needs a long process of continuous support and cultivation. Schools have not traditionally focused on developing creativity. In fact, schooling, as it has historically operated, hurts the development of creativity through standardization, centralization, and assessment. In the future, given the sophistication of learning technologies, schools need to change into a global learning ecosystem in which students have a strength-based, personalized creative learning experience.

References

Australian Curriculum Assessment and Reporting Authority. (2010). A curriculum for all young Australians. Retrieved from http://www.acara.edu.au/verve/_resources/Information_Sheet_A_curriculum_for_all_young_Australians.pdf

Bonawitza, E., Shaftob, P., Gweonc, H., Goodmand, N. D., Spelkee, E., & Schulzc, L. (2011). The double-edged sword of pedagogy: Instruction limits spontaneous exploration and discovery. *Cognition, 120*(3), 322–330.

Common Core State Standards Initiative. (2011). Common Core State Standards Initiative. Retrieved from http://www.corestandards.org/

Council of Chief State School Officers, & National Governors Association Center for Best Practices (NGA Center). (2010, June 2, 2010). Introduction to the Common Core State Standards. Retrieved from http://www.corestandards.org/assets/ccssi-introduction.pdf

Emler, T. E., Zhao, Y., Deng, J., Yin, D., & Wang, Y. (2019). Side Effects of Large-Scale Assessments in Education. *ECNU Review of Education, 2*(3), 279–296.

Kaufman, J. C., & Beghetto, R. A. (2009). Beyond Big and Little: The Four C Model of Creativity. *Review of General Psychology, 13*(1), 1–12.

Lewontin, R. (2001). *The Triple Helix: Gene, Organism, and Environment.* Cambridge, MA: Harvard University Press.

Nichols, S. L., & Berliner, D. C. (2007). *Collateral Damage: How High-Stakes Testing Corrupts America's Schools.* Cambridge, MA: Harvard Education Press.

OECD. (2019, April). PISA 2021 Creative Thinking Framework. *OECD.* Retrieved from https://www.oecd.org/pisa/publications/PISA-2021-creative-thinking-framework.pdf

Partnership for 21st Century Skills. (2007). Framework for 21st Century Learning. Retrieved from http://www.21stcenturyskills.org/documents/frameworkflyer_072307.pdf

Rideout, V., & Robb, M. B. (2019). *The Common Sense census: Media use by tweens and teens, 2019*. San Francisco, CA: Common Sense Media.
Ridley, M. (2003). *Nature via nurture: genes, experience, and what makes us human* (1st ed.). New York, N.Y.: HarperCollins.
Tienken, C. H., & Zhao, Y. (2013). How Common Standards and Standardized Testing Widen the Opportunity Gap. In P. L. Carter & K. G. Welner (Eds.), *Closing the Opportunity Gap: What America Must Do to Give Every Child an Even Chance* (pp. 113–122). New York: Oxford University Press.
Wehmeyer, M., & Zhao, Y. (2020). *Teaching Students to Become Self-Determined Learners*. Alexandria, VA: ASCD.
Zhao, Y. (2012). *World Class Learners: Educating Creative and Entrepreneurial Students*. Thousand Oaks, CA: Corwin.
Zhao, Y. (2016). From Deficiency to Strength: Shifting the Mindset about Education Inequality. *Journal of Social Issues, 72*(4), 716–735.
Zhao, Y. (2018a). *Reach for Greatness: Personalizable Education for All Children*. Thousand Oaks, CA: Corwin.
Zhao, Y. (2018b). *What Works May Hurt: Side Effects in Education*. New York: Teachers College Press.
Zhao, Y. (2020a). Another education war? The coming debates over social and emotional learning. *Phi Delta Kappan, 101*(8), 42–48.
Zhao, Y. (2020b). Two decades of havoc: A synthesis of criticism against PISA. *Journal of Educational Change*, 1–22. https://doi.org/10.1007/s10833-019-09367-x
Zhao, Y., & Gearin, B. (2016). Squeezed Out. In *Creative intelligence in the 21st century* (pp. 121–138): Springer.

Yong Zhao is a Foundation Distinguished Professor in the School of Education at the University of Kansas and a professor in Educational Leadership at the Melbourne Graduate School of Education in Australia. He is an elected member of the National Academy of Education and a fellow of the International Academy of Education.

Chapter 14
Engaging Uncertainty: Principles and Provocations for Promoting Creative Learning Futures

Ronald A. Beghetto

Abstract How we might think about the future of creativity, learning and technology? There are, of course, numerous ways to approach this question. One way would be to systematically work through the three concepts and consider possible futures for each. Another approach would be to work within some combination of the relationship among these concepts in relation to a future orientation. Any of these approaches have the potential to result in interesting conjectures for considering the future possibilities for creativity, technology, and learning. Regardless of the approach one takes, however, it is important to recognize that uncertainty plays a central role in any discussion of the future. The purpose of this chapter is to explore how we might move toward creative learning in the context of uncertain futures and consider the potential role that technologies play in these futures. More specifically, I discuss key principles of creative learning under uncertainty and invite researchers and practitioners to consider how existing and new technologies might help or further impede efforts aimed at promoting creative learning in and beyond the walls of schools and classrooms. I organize the discussion in this chapter around a set of provocations aimed at considering the role technology might play in supporting each phase of creative learning now and into the future.

Keywords Creative learning · Uncertainty · Technology · Learning futures · Education

How might we think about the future of creativity, technology and learning? A first step in addressing this question is to briefly define creativity, learning, and technology. Creativity refers to thoughts, actions or experiences that are judged as both novel and meaningful as defined in particular context and a given individual or social audience (Kaufman and Beghetto 2009; Plucker et al. 2004). Learning refers to a relatively stable change in understanding and behavior (Alexander et al. 2009).

R. A. Beghetto (✉)
Mary Lou Fulton Teachers College, Arizona State University, Tempe, AZ, USA
e-mail: Ronald.Beghetto@asu.edu

D. Henriksen, P. Mishra (eds.), *Creative Provocations: Speculations on the Future of Creativity, Technology & Learning*, Creativity Theory and Action in Education 7,
https://doi.org/10.1007/978-3-031-14549-0_14

Technology refers here to any digital or non-digital tool that can support (or inadvertently impede) creativity and learning. The primary focus in this chapter will be on the potential use of digital technologies in support of creative sense making during the process of learning.

Having now briefly defined these concepts we can now explore various possibilities in exploring the question of how we might think about the future of creativity, learning and technology? There are, of course, numerous ways to approach this question. One way would be to systematically work through the three concepts and consider possible futures for each. Another approach would be to work within some combination of the relationship among these concepts in relation to a future orientation. Any of these approaches have the potential to result in interesting conjectures for considering the future possibilities for creativity, technology, and learning. Regardless of the approach one takes, however, it is important to recognize that uncertainty plays a central role in any discussion of the future.

Uncertainty also plays an important role in present efforts aimed at change and transformation. This is because the near and distant future always has some level of unknowability. Uncertainty, however, also serves another role when it comes to learning, creativity and change. Encounters with uncertainty are also an opportunity to move into new ways of thinking and acting. In this way, uncertainty serves as a catalyst and condition for creative learning now and into the future (Beghetto 2020b). Uncertainty can, of course, be an overwhelming experience and result in us attempting to avoid it and maintain the status quo or attempt to quickly resolve it and even defer our creative agency to others (Beghetto 2021). How we experience and respond to uncertainty is thereby contingent on a host of factors and can result in a variety of intended and unintended outcomes, some of which may lead to creative learning and others which may reinforce existing knowledge and practices.

Technologies can accelerate or amplify these experiences and outcomes or even become an outcome in and of themselves. Consider the profound uncertainties generated by the COVID-19 pandemic. Multitudes of students and teachers found themselves rapidly propelled into online learning experiences. This was a new experience for many teachers and students. Moreover, it was the kind of large-scale change that did not afford the typical, protracted considerations of potential benefits or hazards of moving to online learning (which likely would have occurred if the world was not facing a pandemic). This rapid shift to a fully digitally mediated learning experience accelerated a search for an immediate, viable solution to continue carrying out the project of educating students. Many educational institutions turned to video-based, multi-participant platforms such as *Zoom*, *Google Meet*, and *Microsoft Teams* coupled with learning management systems to provide a blend of live and asynchronous teaching and learning experiences. This shift was both rapid and large scale. Can we view this shift as a sign or omen of our teaching and learning futures? Do such changes promote creative learning? At best we can say, it depends. This is because changing the medium of teaching and learning does not (of course) result in more creative instructional practices or learning experiences.

Although it is true that it took a pandemic to shift much of the world to consider and adopt (at least temporarily) a new modality of teaching and learning, I would argue that it is also true to say that the shift didn't necessarily accomplish much by way of changing the goals or approach to teaching and learning in that new digital modality. Perhaps the need to rapidly resolve the uncertainty of simply how to carry on with existing approaches to teaching and learning was too great and thereby came at the cost of leveraging the uncertainty to also transform how teaching and learning might be approached differently? A pandemic is not necessary for such a shift to occur. Rather, what seems to be needed is a new set of principles that offer a feasible alternative for making small, but potentially transformative shifts in the how and why we engage in the process of teaching and learning.

Although we may be tempted to believe that new, creative learning futures can only be realized through the assistance of new technologies that transform or disrupt current educational practices, it is important to realize that it is less about developing and adopting new technologies and more about adopting new principles of creative learning. Indeed, even the newest of technologies when used in conjunction with existing assumptions and principles of teaching and learning often do little more than offer new modalities for engaging in the same prototypical practices.

In this essay, I will approach the question of how we might move toward creative learning futures and consider the potential role that technologies might play in these futures. More specifically, I will introduce key principles of creative learning in an effort to invite researchers and practitioners to consider how existing and new technologies might help or further impede efforts aimed at promoting creative learning in and beyond the walls of schools and classrooms. I open the essay by describing the key principles of the concept of creative learning. I then offer provocations aimed at considering the role technology might play in supporting each phase of creative learning now and into the future.

14.1 Creative Learning: Principles and Provocations

Creative learning refers to a blend of individual and social processes that result in new and meaningful understanding for oneself and others (Beghetto 2016). The concept of creative learning highlighted herein is based on a program of research (e.g., Beghetto and Schreiber 2017; Beghetto and Schuh 2020; Beghetto and Yoon 2021; Gajda et al. 2017) aimed at clarifying how educators can design learning experiences, which have the goal of promoting new and meaningful learning in and beyond the immediate learning event.

Although it is beyond the scope of this essay to provide a detailed overview of creative learning, it is important to highlight a few core features and principles of creative learning as they will set the stage for considering possible educational futures aimed at supporting these efforts (See Fig. 14.1).

As depicted in Fig. 14.1, creative learning commences with and continues from encounters with uncertainty, which is structured by a series of creative sense-making processes and experiential events. This sense making process can occur at the individual and social level. It can also occur continuously across time and space.

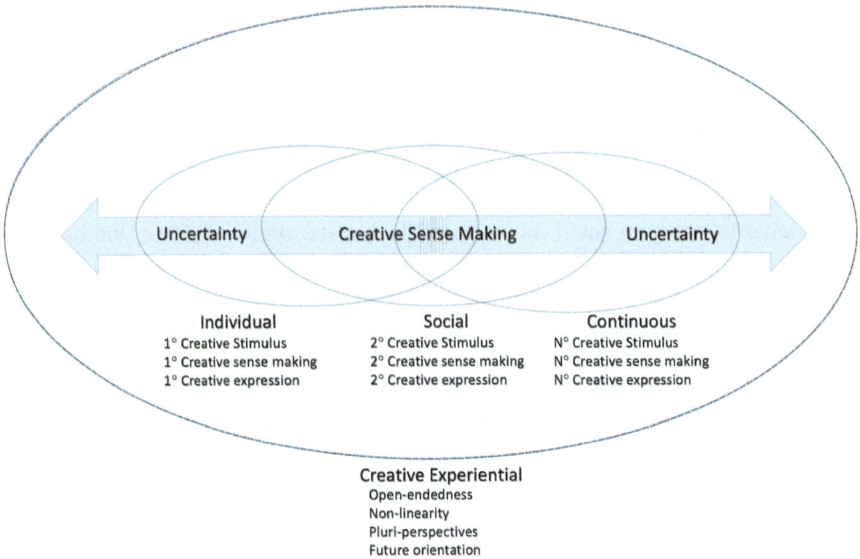

Fig. 14.1 Core features of creative learning

As will be discussed digital technologies can play an important role in the context of a formal educational context by supporting the continuous expression of creative learning across time and space.

In what follows, I outline creative learning principles and provocations, aimed at helping researchers and educators consider how creative learning might be supported (rather than inadvertently suppressed) by current and existing technologies. More specifically, I will work through the four interrelated spheres of creative learning experiences depicted in Fig. 14.1 (i.e., subjective, social, continual, and experiential) - outlining the principles associated with each of these spheres and offer provocations for considering the role that technologies might play now and into the future.

14.1.1 Individual Principles

At the individual level, creative learning commences with an encounter with uncertainty in the form of a creative learning stimulus. *Creative learning stimuli* typically are conceptualized as pertaining to discrepant academic subject matter but can emerge from most any planned (or unplanned) discrepancies between what students already know and what they are encountering in the context of a learning event (Beghetto and Schuh 2020; Beghetto and Vasquez in press). What distinguishes creative learning stimuli is that they are imbued with uncertainty, such that learners are provoked to engage in a process of creative sense making. In this way, creative learning stimuli have been described as optimally discrepant (Beghetto 2016) in that they provoke a sense of doubt and wonder (Glăveanu 2020), but are not too discrepant to be considered unknowable.

Students' are thereby motivated to engage with the creative learning stimulus in an effort to resolve it through a process of *creative sense-making*. Creative sense making refers to a creative combinatorial process (Ward and Kolomyts 2010) that involves developing a new and personally meaningful understanding (Beghetto and Kaufman 2007) through the blending of one's own prior understanding (Beghetto and Schuh 2020; Schuh 2017) and the novel stimuli. Novel stimuli can take almost any form from new ideas (e.g., a new way of thinking about literature), physical artifacts (e.g., beakers, test-tubes, and other scientific objects in a science lab) and digital representations (e.g., tessellations and simulations in a mathematics course). Although creative sense-making typically is considered to be an intra-psychological process, it is not sealed-off from the socio-cultural and material resources of the learning event (Beghetto and Vasquez in press; Glăveanu et al. 2019; Vasquez et al. 2021). Indeed, the learner is never truly alone in the process of creative meaning making as students are always and already in dialogue with actual and imagined interlocutors and audiences in the process of creative sensemaking (Beghetto 2017; Beghetto and Schuh 2020; Glăveanu et al. 2019; Wegerif 2008).

Once learners are able to develop a new and personally meaningful understanding, they can then be said to have experienced creative learning at the subjective, primary, or mini-c level (Beghetto and Kaufman 2007; Runco 1996; Runco and Beghetto 2019; Stein 1953). This is depicted in Fig. 14.1 as the first oval.

14.1.2 Individual Technological Provocations

The first phase of creative learning focuses on learners' subjective encounters with uncertainty in the form of creative learning stimuli, which is then followed by an effort to engage in creative sensemaking. When successful, the outcome of this phase is the development of a new and personally meaningful understanding of learning stimuli. Educators and researchers are therefore invited to speculate on how current and future technologies can play a role in supporting this phase of the process. The following are a set of provocations aimed at generating possibilities for considering the role of technology in this capacity.

- *How might digital technologies be used to provoke creative learning?* Although technology already plays a key role in this regard, the question for the future will be to what extent will technology be more prominently featured in formal learning environments. Indeed, current and future technologies hold much potential for supporting creative learning at the individual level by introducing learners to a nearly endless array of novel learning stimuli, which can result in new and personally meaningful understandings. Although students use technologies to provoke creative learning (e.g., students' using personal technological devices to access an array of digital content and experiences), it is still the case that in many formal learning environments educators rely heavily on third-party curated (e.g., textbooks) and teacher curated (e.g., supplementary videos, websites, etc.) learning stimuli for the purpose of having students converge on predetermined ways of knowing and understanding. Creative learning can still occur in such situations, but it is usually an unexpected or ancillary outcome rather than the primary goal (Beghetto 2013). How might educators use technologies to promote creative learning in a more planned and pervasive way? One way, along the lines of what proponents of self-directed and personalized learning have argued, is to use technologies to provide students with increased agency and voice in their own learning, including using technologies to access and communicate new understandings (Henriksen et al. 2016; Wehmeyer and Zhao 2020). Building from such suggestions, educators might consider even broader possibilities for planned creative learning experiences by leveraging the ease and access to a more expansive array of creative learning stimuli for students to engage within and beyond the classroom.

- *How might learners use digital technologies to support creative sensemaking?* The role of technology in personal, creative sensemaking is an under-realized area of research and educational practice. Although educators have increasingly used learning management system (LMS) to encourage students to "post" and "reflect" "on their learning, this is typically limited to responding to a set of predetermined, retrospective and often quite general writing prompts that the teacher has generated. How might such efforts be expanded to support creative learning at the individual level? One way would be for educators to use technologies that encourage students to share questions and insights "on the fly" and in the midst of creative learning experiences. Limiting the use of technologies to retrospective reflections seems to be a missed opportunity for realizing the potentialities of current and emerging technologies to support creative learning. Educators might consider how technologies to support what might be called, "creative sensemaking self-dialogues''wherein students have multiple, on-going opportunities to more actively work through and document their creative sensemaking process (starting with encountering a novel learning stimulus and on through the full sensemaking process of those stimuli). This might include everything from audio recordings of their thoughts, using virtual note boards and annotations, or any manner of tools that allow students to externalize and engage with their own thinking and creative sensemaking process. It is worth considering how supporting this early, subjective phase of creative learning might simultaneously support (and potentially impede) the creative learning of individual learners and also help make more transparent and visible the creative sensemaking process for educators and researchers.
- *What role might technologies play in the development of new and personally meaningful understanding of academic subject matter?* New and personally meaningful understandings cannot be separated from socio-cultural and contextual features of past, present, and future learning possibilities that are intertwined with the subjective creative meaning making process (Beghetto 2020a; Glăveanu

2021; Schuh 2017). With respect to potential implications for technology, the technological tools that have been used in support of provoking and supporting creative sensemaking might also play a role in this phase of the process. Examples might include, using such tools to help make the features of the sensemaking process more visible (both during the immediate learning event and for future learning events). This, in turn, can serve as a prompt for helping support learners now and into the future in becoming aware of the various dynamic socio-cultural and contextual factors and tools at play in their own creative sensemaking process. Educators and researchers might then consider the pedagogical and creative implications of having students engage in reflections and prospections guided by more personalized, technological artifacts (rather than more generic, retrospective prompts). Such efforts might also help students more effectively communicate their creative understandings and their creative process for arriving at that understanding to others. This, in turn, can facilitate the broader creative learning process beyond the subjective level and into the social sphere wherein students' unique understandings and processes contribute to the creative learning of others.

14.1.3 Social Principles

The initial or primary phase of creative learning focuses on the social or interpsychological phase of the process (depicted in Fig. 14.1 as the second oval). The social phase mirrors the individual phase in some ways but differs in that the various subjective processes are externalized in the social context. More specifically, the social phase starts with an opportunity for learners to share their own unique and personally meaningful conceptions with peers and teachers. If students share unique conceptions that introduce uncertainty into the learning context, those conceptions serve as secondary creative learning stimuli for their peers and teachers (Runco and Beghetto 2019). In this way, uncertainty is reintroduced into the learning environment in the form of a student generated creative learning stimulus and animates the possibility of creative learning for others.

In order for the creative learning process to move forward at the social phase, however, peers and teachers need to be willing to engage in (rather than dismiss) the uncertainty introduced and, in turn, work through their own collective, creative sense-making process. Although individuals may still be engaging in subjective and internalized sensemaking processes during this phase, there is also opportunities for members of the learning community to externalize their thinking and work collectively to discuss and make sense of the uncertainties provoked by the students' (and teachers') unique perspectives. It is in this way that primary creative learning experiences can make secondary creative contributions to others in the immediate learning environment – moving from subjective, mini-c creative insights to inter-subjective little-c creative contributions (Beghetto 2007; Runco and Beghetto 2019).

14.1.4 Social Technological Provocations

Given that the secondary phase of creative learning focuses on social creativity, which commences with members of the immediate digital or physical learning environment engaging with individual students' creative understandings, what role might technology play in supporting this process? As with the primary phase of creative learning, the social phase requires that members of the learning environment are willing to engage with and attempt to make sense of creative learning stimuli encountered when attempting to understand the creative ideas and insights of others. When successful, the outcome of this phase is the development of new and personally meaningful understandings amongst members of the learning community. The following are a set of provocations aimed at generating possibilities for considering the role of technology at the secondary phase of creative learning.

- *How might technologies be used to help individuals communicate creative insights and outcomes to peers, teachers and others in their social environment?* The way in which creative learning typically moves from the individual to the social is from students having an opportunity and willingness to share their unique ideas. This phase of creative learning can be thought of as moving between intrapersonal to interpersonal creative ideation (Beghetto 2007). How might technologies play a role in expanding the opportunities for creative expression to move from the individual to the social phase of creative learning? This is a key question for educators and researchers to consider with respect to current and future technologies. The use of digitally mediated discussions (e.g., LMS discussion boards) is one way that this can (and does occur), but there likely are many additional opportunities for both small and large group exploration of the creative insights of individual students. Teachers who are equipped with an understanding of the full process of creative learning might more intentionally and actively consider ways for students to not only share their perspectives (e.g., video, audio, and other multi-model technology supported forms of expression), but also for these perspectives to be engaged with by teachers and students. As discussed in the previous section, in addition to simply having students describe their insights and ideas, providing opportunities for students to curate, document, and share artifacts of the process they used to arrive at such ideas might also be useful in prompting the social creative learning phase. Current and future technologies can play a key role in this form of ongoing curation and social engagement with individual creative learning outcomes.
- *How might members of the immediate physical or digital learning environment use technologies to support creative sense making?* As with individual creative learning, sense making plays a central role in moving toward a new and personally meaningful understanding at the social level. Leveraging technologies that engage varied perspectives and insights of the social environment can serve to support creative sensemaking of students who share their ideas with others (e.g., helping them test out and refine their perspectives, receive feedback from multiple participants) and also help promote engagement with the uncertainty inherent

in novel understandings. Indeed, if students and teachers in the immediate social environment are expected and willing to engage with (rather than simply dismiss) novel conceptions of peers, then it is possible that they can support creative learning in the immediate learning environment (Beghetto 2013). Technologies can play a key role in providing multiple forms of engagement (beyond group discussion) for individuals, small groups, and the larger group to attempt to make sense of conceptions and perspectives different from their own and thereby carry creative learning forward. Students might for example share or present digital creative learning artifacts of their own sense making process with each other in an effort to provoke creative learning at the social level. Such artifacts can be shared in the physical learning environment or housed in LMS or other digital platforms for ongoing asynchronous engagement.

- *What role might digital technologies play in the development of new and personally meaningful understandings amongst social groups?* As with individual creative learning, unless there is a way to document and curate the creative insights and ideas that occur in the social environment, they are likely to be lost or, at best, carried forward in fragmentary memories and recollections of the creative learning event. It is here that technologies can play an increasingly important role in not only establishing a record of the social creative learning event, but also in keeping the creative potential of social creative learning artifacts alive for immediate participants and future audiences beyond temporally constrained learning environments and events. Even somewhat basic and ubiquitous technologies (e.g., video recordings, digital images of artifacts) can be used toward this end, but it is also important to consider what kinds of current or new technologies might be used or developed for this specific purpose.

14.1.5 Continuous Principles

The most important aspect of creative learning with respect to supporting learning futures is that it has the potential to continue on indefinitely and contribute to the creative sensemaking of others beyond the immediate learning event (Beghetto and Vasquez in press). This process is depicted in the third oval, wherein external and future audiences can engage with reflections, representations, and artifacts from primary and secondary creative learning experiences. There are several ways this can occur. One way is if participants in the initial creative learning event later reflect on a particular creative learning experience, which in turn can serve as a provocation of uncertainty for new learning for oneself and others. These kinds of reflective experiences likely already occur for students and teachers in formal and informal ways, such as when teachers reflect on a taught lesson or when planning new lessons. This can also occur when students reflect on previous lessons. To the extent that such individual reflections serve as creative learning stimuli and then new subjective or social creative learning processes can commence.

Another way this can occur is if teachers or students draw on and share past creative learning experiences in future social or professional situations. Examples of this can include teachers and students participating in a professional development experience, presenting at a conference, or recollecting experiences in some other social venue. Yet another way this can occur is if artifacts of creative learning are collected, curated and shared. Developing and curating creative artifacts of individual and social creative learning are perhaps the most reliable way to maintain the indefinite, continuous potentiality of provoking future creative learning.

An example of the full trajectory of creative learning from individual to continuous creative learning is reported in a recent dissertation study (Han 2019). Han (2019) documented the development and impact of a creative learning artifact developed by Jinwoo, a fourth grade bilingual student from Korea, who created an illustrated book written in Korean and English aimed at documenting and sharing his personally meaningful insights, experiences and emotional reactions to the unexpected events and activities of elementary schooling in the United States. The book represents an example of a generative and powerful creative artifact that was initially developed by the student as an expression of individual creative learning, which was later shared with and provoked social creative learning amongst students and educators in the elementary school context. Given that the artifact was digitized and shared through a dissertation study, it has now moved beyond the immediate individual and social context and has become a potential creative learning stimulus that can continuously contribute to the creative learning of future educators, students and scholars.

14.1.6 Continuous Technological Provocations

As mentioned, when it comes to creative learning futures, it is the third phase of creative learning that arguably holds the most technological promise and potential for animating and sustaining continuous creative contributions. Indeed, creative learning (much like all learning) in educational environments tends to be bounded by a particular time and space. Once the creative learning event has passed, and the learners leave the setting, there is little record or artifact of what has occurred beyond participants' recollections and perhaps any assignments or grading records, which may be kept by students and teachers. The following provocations are aimed at spurring new thoughts and actions regarding the role technologies might play in going beyond the typical fragmented documentation and curation of creative learning and toward supporting the continuous phase of creative learning.

- *How might digital technologies be used to document and produce creative artifacts?* Artifacts typically are thought of as frozen features of a past event, not necessarily as having a future orientation. When it comes to continuous creative learning, however, creative artifacts hold the ever-present potential to carry new learning forward with future audiences of learners, educators and scholars. Creative learning artifacts can include any manner of elements generated prior to, during and following creative learning endeavors (e.g., lesson plans, project overviews, videos of key events, digitized creative learning products, audio-visual and text reflections). Current and future technologies can play an important role in documenting and producing artifacts, not for archiving or filing them away but rather for sharing and communicating creative learning events that can continuously contribute to creative learning futures. One way this can occur is if educators and researchers more actively recognize the importance of using technologies for the purpose of carrying creative learning forward and anticipating this need in designing and carrying out learning experiences. This is arguably one of the most important ways of ensuring creative learning futures that build on and transforms the learning of the past.

- *What role can technologies play in supporting the curation and access of creative learning artifacts?* In order for technologies to play a role in propelling creative learning forward, it is important for educators and researchers to consider how they might make creative learning artifacts more easily available and accessible. Fortunately, providing ease of communication and access to learning artifacts is a unique feature of learning technologies and how they can be used in support of creativity (Henriksen et al. 2016). Consequently, using existing and developing new technologies to curate learning events so they can be meaningfully labeled, searched, and accessed by future audiences of learners and educators is an important first step for educators and researchers interested in making continuous creative learning contributions to others. Moreover, it is also important to recognize and plan for curation at all stages of the creative learning process. This can be facilitated by the use of existing technologies, such as: living, cloud-based documents (e.g., Google Docs, Dropbox Paper, blogs); project websites depositories that provide links to on-going and completed creative learning projects; and even the development of online creative learning clearinghouses where educators and learners can submit creative learning artifacts. Establishing sharing agreements, permissions, and other safeguards (e.g., intellectual property and privacy safeguards) will be an important consideration in sharing any educational artifacts. Educators and researchers will therefore need to consider how they can strike a meaningful balance between permissions and consents with accessibility and access for future audiences. Although this may present a challenge, it is not insurmountable as there are examples of online data (e.g., https://www.apa.org/research/responsible/data-links) and project repositories (https://projectbbq.com/schools/Iowa_BIG_Newbo/public-pool) that can serve as an example and starting point for curating and sharing creative learning artifacts.

14.1.7 Experiential Principles

Finally, it is also important to recognize that creative learning represents a specific type of a broader conceptualization of creative experiences (Glăveanu and Beghetto 2020). Creative experiences represent "*novel person–world encounters grounded in meaningful actions and interactions, which are marked by the principles of: open-endedness, nonlinearity, pluri-perspectives and future-orientation*" (Glăveanu and Beghetto 2020, p. 2). Recognizing how creative learning represents a specific type of creative experience requires a bit of unpacking in order to see the connections and understand how the experiential features connect to creative learning.

First, the full expression of creative learning represents a novel person-world encounter because it entails people engaging in meaning making actions and interactions, which are prompted and sustained by encounters with novel learning stimuli. Indeed, creative learning in its full expression involves a series of continual actions and interactions with stimuli and participants in (and beyond) the learning environment. As has been discussed, creative sensemaking does not occur in the isolation of solitary minds, but rather is distributed and animated through the active expression and interactions amongst participants in and beyond a particular creative learning event.

In this way there is always an *open-endedness* to creative learning. Indeed, open-endedness refers to the action orientations taken-up by participants in the creative learning experience and is represented in their willingness to "explore emerging directions" and "go off script" -- moving away from "predetermined [learning] outcomes" and toward exploring the creative potential of the "unexpected" and "different views of others" (Glăveanu and Beghetto 2020, p. 2). This open-endedness of creative learning is also reflected in participants' willingness and attunement to creative ruptures (Beghetto 2020b) and micro-serendipities (Ross and Vallée-Tourangeau 2020) that emerge when students encounter, share, and engage with discrepant learning stimuli.

The *open-endedness* of creative learning is also reflected in the continuous and inconclusive potential of creative learning outcomes (Corazza 2016). The documentation, curation and exhibition of creative learning events opens-up ever expansive vistas of creative possibilities in how others might engage with and learn from creative learning events (Vasquez et al. 2021). Indeed, there are manifold trajectories of creative learning that can occur beyond the temporal and situational confines of creative learning events. A consideration for educators and researchers designing creative learning experiences is to think about how they might plan with the principle of open-endedness as a way of helping to facilitate, notice, and cultivate creative learning for individuals, classroom groups, and beyond.

Creative learning also reflects the *non-linearity* of creative experiences. Although the processes depicted in Fig. 14.1 represent a seemingly linear trajectory; in practice, creative learning involve recursive and bi-directional features that dynamically transform creative learning outcomes. The interactions, sensemaking, and experiences that occur in creative learning have a give-and-take and even messiness to the process (Tanggaard and Juelsbo 2016). This is because working through the uncertainty of creative learning experiences have an emergent and unpredictable quality to them (Sawyer 2012).

The unpredictable nature and messiness inherent in creative learning can occasionally be experienced by students and teachers as frustrating and stifling. It is therefore important to recognize that although uncertainty is filled with possibilities, it can also contain setbacks, wrong turns, failures, and frustrations. This requires thinking about how uncertainty might be structured to reduce feelings of being at a loss for how to proceed (Beghetto 2019) as well as rethinking the role of failure in creative work (Manalo and Kapur 2018; von Thienen et al. 2017). Consequently, it is important to anticipate and plan for the kinds of supports that might be necessary to help students and teachers navigate the setbacks and impasses inherent in the creative learning process.

Plurality of perspectives is also central to creative learning experiences. Creative learning occurs in the context of difference (Glăveanu and Beghetto 2016), including: differences in prior knowledge and encounters with new learning stimuli (Schuh 2017), differences in how stimuli are encountered and understood amongst participants in the learning event, and differences in perspectives beyond the immediate learning event. The sensemaking processes featured in this conception of creative learning draws on the somewhat idiosyncratic and subjective aspects of creative, combinatorial processes (Rothenberg 2015; Ward and Kolomyts 2010). These creative sensemaking processes, however, are not impermeable or isolated from other participants (past, present, and future) or the socio-cultural and material entanglements of the creative learning event (Glăveanu et al. 2019).

Creative sense making occurs in the context of dynamic dialogical and socio-material arrangements and perspectives afforded by people, ideas, objects, and experiences in the immediate environment and with past, present and imagined future interlocutors and encounters (Beghetto 2017; Malafouris 2014, 2020; Wendy Ross and Vallée-Tourangeau 2018; Tanggaard and Beghetto 2015). Although the socio and material context is sometimes overlooked (van der Zanden et al. 2020), it looms large in the creative learning process. Indeed, the sociocultural features become animated in the socio-material interactions of creative learning and help shape participants' emerging perspectives and interpretations of the events and outcomes of the creative learning process. This includes technological resources, which can facilitate (or inadvertently impede) the creation and sharing of creative ideas and artifacts (Henriksen et al. 2016).

The dynamic, emergent and inconclusive nature of creative learning thereby always and already assumes a *future orientation.* What has and is occurring in creative learning events shape new trajectories of possibilities. These new possibilities can be realized through retrospection (e.g., reflecting on prior creative learning events can engender new learnings) and through prospection by considering future "as if" and not yet experienced states of teaching and learning (Glăveanu et al. 2019; Greene 1995; Warnock 1978). In this way, there is an inherent generative potential in creative learning that can result in continuous creative contributions to the learning of others in and beyond any immediate creative learning event.

14.1.8 Experiential Technological Provocations

- *How might technologies play a role in supporting the design, experience, and continuation of creative learning?* The principles of creative learning experiences briefly outlined above can serve as a lens for thinking about how technologies can be used to design and support the full expression of creative learning from the subjective experience on through continuous creative learning contributions beyond the learning event. Educators and researchers might consider how technologies offer opportunities for creative learning at any stage to maintain a quality of open-endedness by allowing students and educators to continually revisit, revise, and engage with creative learning stimuli. Educators often use digitally mediated discussions and reflections for such purposes, but those activities usually have a preset due date (or expiration date). What if educators used digital forums and discussion boards as a continuous feature of creative learning during the immediate subjective and social phases of creative learning and also as a living and dynamic record in support of continuous creative learning? Doing so might help to animate and infuse other facets of creative learning experiences, such as non-linearity because they enable current and future participants to literally go back and forth in time and revisit earlier stages of creative learning and carry it forward into the future. Moreover, technologies can also help facilitate the plurality of perspectives by breaking down typical physical barriers of classroom learning by inviting feedback and perspectives of multiple audiences during the planning of creative learning events, the subjective and social phases of creative learning events, and when audiences respond to creative learning artifacts. Using technologies to animate open-endedness, nonlinearity, and pluri-perspectives at all phases of creative learning will also instantiate the future-orientation of creative learning by maintaining the possibility for continuous, on-going creative contributions.

14.2 Concluding Thoughts

As has been discussed in this essay, thinking about and working toward new creative learning futures will, at some point, run up against deep uncertainties. As I've attempted to highlight in this essay, the experience of uncertainty, albeit uncomfortable, is also an opportunity to think and act in new ways. This essay and the other essays in this volume provide an invitation for readers to engage in fresh thinking, speculation, and new possibilities for understanding the future of learning, creativity, and technology. In this essay, I focused on introducing the concept of creative learning and how we have an opportunity to leverage existing and new technologies to move creative learning beyond the walls of schools and classrooms. Such efforts entail persistent engagement with uncertainties. Indeed, as has been discussed, engaging with uncertainty is inherent in the on-going and indefinite process of creative sense-making.

Current and future technologies likely will play a non-trivial role in helping to structure students' and teachers' creative learning encounters with uncertainty by offering opportunities to curate creative learning stimuli, support creative sense-making, and continuously contribute new insights to others through open-ended, non-linear, pluri-perspectival and ultimately future-oriented learning experiences. Although creative learning connotes something new and, maybe even, something quite radical when considered in light of the seemingly fixed, pedagogies of sameness that characterize prototypical educational experiences (Beghetto and Yoon 2021; Glăveanu and Beghetto 2016), in reality, educational philosophers have long recognized that education has at its core a future orientation aimed at releasing the imagination (Greene 1995) and realizing new and more promising possible futures (Dewey 1910; Vygotsky and Sharpe 2004; Warnock 1978).

Education in practice, however, has long been encumbered by pedagogies that perpetuate a state of knowable immediacy, which reaches to a near past for guidance on what should be done now in order to prepare young people for a foreseeable future. Such an approach comes with the benefit of reducing the discomfort of present and future uncertainties but has the drawback of constraining our collective imagination as to how we might enact new and bold present possibilities for moving toward unknowable futures. Indeed, it is the uncertainty of unknowable futures that offers actionable potential for realizing more promising creative learning and technological futures. If we are willing to shift our thinking from the fixed and foreseeable and toward the inconclusive and unknowable. Indeed, the full expression of creative learning requires that we provide a means for artifacts of those experiences to be carried into the future, not fossilized and filed away. When education is viewed from a creative learning vantage point, technologies play an entirely different role than how they are typically used in education. The use of digital technologies in education is thereby released from the limiting constraints of simply trying to provide a modality for attempting to replicate what traditionally has occurred in schools and classrooms and moves us toward imagining new possibilities for how technologies can be used to carry creative learning forward into new and more promising futures.

References

Alexander, P. A., Schallert, D. L., & Reynolds, R. E. (2009). What is learning anyway? A topographical perspective considered. *Educational Psychologist, 44*(3), 176–192. https://doi.org/10.1080/00461520903029006

Beghetto, R. A. (2007). Ideational code-switching: Walking the talk about supporting student creativity in the classroom. *Roeper Review, 29*, 265–270.

Beghetto, R. A. (2013). *Killing ideas softly? The promise and perils of creativity in the classroom.* Information Age Publishing.

Beghetto, R. A. (2016). Creative learning: A fresh look. *Journal of Cognitive Education and Psychology, 15*, 6–23.
Beghetto, R. A. (2017). Creative openings in the social interactions of teaching. *Creativity: Theories-Research-Applications, 3*, 261–273.
Beghetto, R. A. (2018). *Beautiful risks: Having the courage to teach and learn with creativity.* Lanham, MD: Rowman & Littlefield
Beghetto, R. A. (2019). Structured Uncertainty: How Creativity Thrives Under Constraints and Uncertainty. In Carol A. Mullen (Ed.), *Creativity Under Duress in Education?* (Vol. 3, pp. 27–40). Springer International Publishing. https://doi.org/10.1007/978-3-319-90272-2_2
Beghetto, R. A. (2020a). Creative Learning and the Possible. In V. P. Glăveanu (Ed.), *The Palgrave Encyclopedia of the Possible* (pp. 1–8). Springer International Publishing. https://doi.org/10.1007/978-3-319-98390-5_57-1
Beghetto, R. A. (2020b). Uncertainty. In V. P. Glăveanu (Ed.), *The Palgrave Encyclopedia of the Possible* (pp. 1–7). Springer International Publishing. https://doi.org/10.1007/978-3-319-98390-5_122-1
Beghetto, R. A. (2021). How Times of Crisis Serve as a Catalyst for Creative Action: An Agentic Perspective. *Frontiers in Psychology, 11*, 600685. https://doi.org/10.3389/fpsyg.2020.600685
Beghetto, R. A., & Kaufman, J. C. (2007). Toward a broader conception of creativity: A case for mini-c creativity. *Psychology of Aesthetics, Creativity, and the Arts, 1*, 73–79.
Beghetto, R. A., & Schreiber, J. B. (2017). Creativity in Doubt: Toward Understanding What Drives Creativity in Learning. In R. Leikin & B. Sriraman (Eds.), *Creativity and Giftedness* (pp. 147–162). https://doi.org/10.1007/978-3-319-38840-3_10
Beghetto, R. A., & Schuh, K. L. (2020). Exploring the connection between imagination and creativity in academic learning. In D. D. Preiss, D. Cosmelli, & J. C. Kaufman (Eds.), *Creativity and the Wandering Mind* (pp. 249–267). Academic Press. https://doi.org/10.1016/B978-0-12-816400-6.00011-0
Beghetto, R. A., & Vasquez, A. M. (in press). Creative learning: A pedagogical perspective. In L. J. Ball & F. Vallée-Tourangeau (Eds.), *Routledge International Handbook of Creative Cognition*. Routledge.
Beghetto, R. A., & Yoon, S. S. (2021). Change Through Creative Learning: Toward Realizing the Creative Potential of Translanguaging. In C.A. Mullen (Ed.), *Handbook of Social Justice Interventions in Education* (pp. 1–20). Springer International Publishing. https://doi.org/10.1007/978-3-030-29553-0_3-1
Corazza, G. E. (2016). Potential Originality and Effectiveness: The Dynamic Definition of Creativity. *Creativity Research Journal, 28*(3), 258–267. https://doi.org/10.1080/10400419.2016.1195627
Dewey, J. (1910). *How we think.* D.C. Heath & Co.
Gajda, A., Beghetto, R. A., & Karwowski, M. (2017). Exploring creative learning in the classroom: A multi-method approach. *Thinking Skills and Creativity, 24*, 250–267. https://doi.org/10.1016/j.tsc.2017.04.002
Glăveanu, V. P. (2020). *Wonder: The extraordinary power of an ordinary experience*. Bloomsbury Academic.
Glăveanu, V. P. (2021). *The possible: A sociocultural theory.* Oxford University Press.
Glăveanu, V. P., & Beghetto, R. A. (2016). *The difference that makes a creative difference* (R. A. Beghetto & B. Sriraman, Eds.). Springer.
Glăveanu, V. P., & Beghetto, R. A. (2020). Creative Experience: A Non-Standard Definition of Creativity. *Creativity Research Journal*, 1–6. https://doi.org/10.1080/10400419.2020.1827606
Glăveanu, V. P., Hanchett Hanson, M., Baer, J., Barbot, B., Clapp, E. P., Corazza, G. E., & Sternberg, R. J. (2019). Advancing Creativity Theory and Research: A Socio-cultural Manifesto. *The Journal of Creative Behavior*. https://doi.org/10.1002/jocb.395
Greene, M. (1995). *Releasing the imagination: Essays on education, the arts, and social change.* Jossey-Bass.

Han, M. (2019). *Exploring the nexus netween bilingual learning, emotions, and creativity: A case study of a former Korean bilingual student's creative artifact (Book).* Doctoral Dissertations. 2352. https://opencommons.uconn.edu/dissertations/2352

Henriksen, D., Mishra, P., & Fisser, P. (2016). Infusing Creativity and Technology in 21st Century Education: A Systemic View for Change. *Journal of Educational Technology & Society, 19*(3), 27–37.

Kaufman, J. C., & Beghetto, R. A. (2009). Beyond big and little: The four c model of creativity. *Review of General Psychology, 13*(1), 1–12.

Malafouris, L. (2014). Creative thinking: The feeling of and for clay. *Pragmatics & Cognition, 22*(1), 140–158. https://doi.org/10.1075/pc.22.1.08mal

Malafouris, L. (2020). Thinking as "Thinging": Psychology with Things. *Current Directions in Psychological Science, 29*(1), 3–8. https://doi.org/10.1177/0963721419873349

Manalo, E., & Kapur, M. (2018). The role of failure in promoting thinking skills and creativity: New findings and insights about how failure can be beneficial for learning. *Thinking Skills and Creativity, 30*, 1–6. https://doi.org/10.1016/j.tsc.2018.06.001

Plucker, J. A., Beghetto, R. A., & Dow, G. T. (2004). Why isn't creativity more important to educational psychologists? Potentials pitfalls and future directions in creativity research. *Educational Psychologist, 39*(2), 83–96 1. https://doi.org/10.1207/s15326985ep3902_1

Ross, W., & Vallée-Tourangeau, F. (2020). Microserendipity in the Creative Process. *The Journal of Creative Behavior*, jocb.478. https://doi.org/10.1002/jocb.478

Ross, Wendy, & Vallée-Tourangeau, F. (2018). For a Good Poet's Made, as Well as Born: The Relational Ontology of Shakespeare's Genius. *Creativity. Theories – Research - Applications, 5*(1), 26–40. https://doi.org/10.1515/ctra-2018-0002

Rothenberg. (2015). *Flight from wonder: An investigation of scientific creativity*. Oxford University Press.

Runco, M. A. (1996). Personal creativity: Definition and developmental issues. *New Directions in Child Development, 72*, 3–30.

Runco, M. A., & Beghetto, R. A. (2019). Primary and secondary creativity. In *Current Opinion in Behavioral Sciences* (Vol. 27, pp. 7–10).

Sawyer, R. K. (2012). *Explaining creativity: The science of human innovation* (2nd ed.). Oxford University Press.

Schuh, K. L. (2017). *Making Meaning by Making Connections*. https://doi.org/10.1007/978-94-024-0993-2

Stein, M. I. (1953). Creativity and culture. *The Journal of Psychology, 36*, 311–322.

Tanggaard, L., & Beghetto, R. A. (2015). Ideational Pathways: Toward a New Approach for Studying the Life of Ideas. *Creativity. Theories – Research - Applications, 2*(2), 129–144. https://doi.org/10.1515/ctra-2015-0017

Tanggaard, L., & Juelsbo, T. (2016). *Mess* (V. P. Glăveanu, L. Tanggaard, & C. Wegener, Eds.; pp. 78–86). Palgrave.

van der Zanden, P., Meijer, P. C., & Beghetto, R. A. (2020). A review study about creativity in adolescence: Where is the social context? *Thinking Skills and Creativity, 38*, 100702. https://doi.org/10.1016/j.tsc.2020.100702

Vasquez, A. M., Koro, M., & Beghetto, R. A. (2021). Creative talent as emergent event: A neurodiversity perspective. In E. Kuusisto, M. Ubani, P. Nokelainen, & A. Toom (Eds.), *Good teachers for tomorrow's schools: Purpose, values and talents in education.* Brill.

von Thienen, J., Meinel, M., & Corazza, G. E. (2017). A short theory of failure. *Electronic Colloquium on Design Thinking Research, 17*, 1–5.

Vygotsky, L. S., & Sharpe, M. E. (2004). Imagination and creativity in childhood. *Trans.). Journal of Russian and East European Psychology, 42*, 7–97.

Ward, T. B., & Kolomyts, Y. (2010). Cognition and creativity. In J. C. Kaufman & R. J. Sternberg (Eds.), *The Cambridge handbook of creativity*. Cambridge University Press.

Warnock, M. (1978). *Imagination*. University of California Press.

Wegerif, R. (2008). Dialogic or dialectic? The significance of ontological assumptions in research on educational dialogue. *British Educational Research Journal, 34*(3), 347–361. https://doi.org/10.1080/01411920701532228

Wehmeyer, M. L., & Zhao, Y. (2020). *Teaching students to become self-determined learners*. ASCD.

Ronald A. Beghetto is an internationally recognized expert on creative thought and action in educational settings. He holds the Pinnacle West Presidential Chair and serves as a Professor in the Mary Lou Fulton Teachers College at Arizona State University. Dr. Beghetto is Editor for the Journal of Creative Behavior, Co-Editor for Review of Research in Education, and Co-Series Editor for Creative Theory and Action in Education (Springer Books). Dr. Beghetto has served as a creativity advisor for LEGO Foundation and the Cartoon Network. He is a Fellow of the American Psychological Association, the Society for the Psychology of Aesthetics, Creativity and the Arts (Div. 10, APA), and the International Society for the Study of Creativity and Innovation (ISSCI).

Chapter 15
Dialogical Provocations: A Creative Trialogue

Vlad Petre Glăveanu, Giovanni Emanuele Corazza, and Ingunn Johanne Ness

Abstract Future-making is a collective enterprise. Learning and creativity are as much psychological as they are social and cultural phenomena. The creation of new technologies requires division of labour, their use connects us with those around us. And the future of the emerging field of learning, creativity and technology studies is ours to envision and to bring into being. Any speculation about the futures (always in plural) of creativity, technology and learning is therefore dialogical and depends on exchanges between different individuals and groups within society. Our final 'creative provocation', then, is meant to recognise and capitalise on the social roots of learning creatively, with technology, and speculate about these futures in an equally dialogical manner. This exchange brings together three authors with expertise in a range of relevant areas, from engineering and learning to sociocultural studies and the psychology of creativity. Our hope in using this format is that it will not only be more enjoyable and authentic but bring insights that both build on and expand what the present book offers in its rich array of contributions, views, and provocations.

Keywords Creativity · Learning · Technology · Sociocultural theory · Dialogue · Dialogism

Future-making is a collective enterprise. It requires people who imagine, who respond to acts of imagination and who are ready to work together in order to turn

V. P. Glăveanu (✉)
School of Psychology, Dublin City University, Dublin, Ireland
e-mail: vlad.glaveanu@dcu.ie

G. E. Corazza
DEI Department, University of Bologna, Bologna, BO, Italy

Université Paris Cité and Univ Gustave Eiffel, Boulogne-Billancourt, France

I. J. Ness
Centre for the Science of Learning & Technology (SLATE), University of Bergen, Bergen, Norway

D. Henriksen, P. Mishra (eds.), *Creative Provocations: Speculations on the Future of Creativity, Technology & Learning*, Creativity Theory and Action in Education 7,
https://doi.org/10.1007/978-3-031-14549-0_15

imagination into reality. In the area of creativity, learning and technology, this underlining collaborative dimension of future-making is brought even more to the fore. Learning and creativity are as much psychological as they are social and cultural phenomena. The creation of new technologies requires division of labour, their use connects us with those around us. And the future of the three is ours to envision and to bring into being. Any speculation about the futures (always in plural) of creativity, technology and learning is therefore dialogical and depends on exchanges between different individuals and groups within society. Our final 'creative provocation', then, is meant to recognise and capitalise on the social roots of learning creatively, with technology, and speculate about these futures in an equally dialogical manner. This exchange brings together three authors with expertise in a range of relevant areas, from engineering and learning to sociocultural studies and the psychology of creativity. Our hope in using this format is that it will not only be more enjoyable and authentic (for us as discussants and for our readers) but bring insights that both build on and expand what the present book offers in its rich array of contributions, views, and provocations. This final chapter is, therefore, not the 'final word' on creativity, technology and learning; it is, in the good tradition of dialogism, a prelude and an invitation – to engage in future-making in open, creative, and ethical ways.

* * *

Vlad Glăveanu Paraphrasing Hartley (1953) in *The Go-Between*, we can say that 'the future is a different country. They do things differently there'. This applies even more to creativity, technology and learning, one of the most rapidly evolving areas of interdisciplinary interest, an area consistently brought to the fore by the current pandemic. The present volume offers a unique collection of perspectives for all those concerned by this field and its future. Its framing in terms of 'creative provocations' is inspired as it captures the tense interplay between accelerated developments, particularly in educational technologies, and the need for further reflection as to how technology depends on and affects creativity and learning. This book raises important questions in this regard for theoreticians and practitioners alike:

Can robots be legitimately considered 'creative' (Chap. 2) and does AI help or hinder creative learning (Chap. 3)? What is the place of technology (Chap. 5) and creativity (Chap. 13) in today's education? Are we able to build creative ecologies (Chap. 4) or will we fall prey to conceptions about creativity and learning that end up harming their development (Chap. 6)? How do we learn, create, and use technology in music (Chap. 10), fine art (Chap. 7), and sports (Chap. 8)? Most of all, what does it mean to live in an age of uncertainty for teachers and students and can we use this uncertainty to our benefit (Chap. 14)? How reflective are we about the current situation (Chap. 11) and do we pay enough attention to those disadvantaged by it (Chaps. 1 and 9)?

It is this last question that I would like to dwell on, and this is the question of ethics (see also Copeland 2022). Creativity, technology and learning are topics that invite considerable optimism about the human condition and our possibilities of transcending some of its key constraints. Through them, the common narrative goes, we empower learners to live up to their 'potential', we help creators expand their knowledge and networks, we build technologies that make life better and push education into the twenty-first century. A lot of these benefits are real and should not be underestimated. But we need to also consider how each expansion of the possible raises new impossibilities that need to be carefully reflected upon and challenged (Glăveanu 2020a). Celebratory discourses about creativity, learning and technology are not as inclusive as we might think, nor are they addressing everyone. If the pandemic showed us anything in education, it is how socioeconomic differences have a major impact on who has access to technology and can therefore use it to continue studying. And even where technology is present, it doesn't mean that learning and creativity are fostered or recognised. Some students are expected to do well in the new digital spaces being opened, others are meant to struggle, and others yet are excluded from the category of 'creative learners'. How, then, can we mobilize creativity, learning and technology to achieve something close to impossible: more equitable forms or educating and living?

Giovanni Emanuele Corazza When speaking about creativity, technology, and learning, there are two points that are critical in my view. The first is the question about "who" is the agent of our educational concern, while the second is about "what" we feel is important and necessary to learn. Once these questions are clarified, then all discussions about "how" technology can be used to enhance learning and creativity of our agent can follow in a linear way.

The answer to the first question could simply be that the agent of our concern is a human being. But a human being is not a static entity: the very same technology that is discussed in this book has dramatic dynamic effects on the human mind, which turns out to be effectively extended by all the available technological means (Clark and Chalmers 1998), although this might raise debates about the authenticity of such extension (Coin and Dubljević 2021). Put simply, the part of the world that is connected to the Internet and that accesses it on a daily basis through personal devices such as a smartphone, is already today a population of cyber-humans, a population that could not live its normal life if it were stripped of the technology it depends upon, and whose characteristics and intellectual capacities are destined to evolve very significantly within the span of a single generation.

If the above statement is accepted, then it follows immediately that the design of learning experiences that exploit new technologies (e.g., Chaps. 6 and 9) are not only useful but "natural" for cyber-humans. The sanitary emergency caused by the COVID-19 pandemic has given a significant push towards the online extension of our lives, that should not and will never completely replace our physical endeavors, but that will certainly affect them. A restoration of pre-pandemic conditions would appear to be impossible. In this view, the very interesting discussions about "creativity of humans vs. creativity of machines" (e.g., Chaps. 2, 3, 7 and 10) are still affected by a sort of human vs. machine duality, that is actually being gradually replaced by the finely nuanced hybridization of the cyber-human being.

The second question addresses "what" we should develop through learning in our population, and many of us point towards creativity as a fundamental target

(Chaps. 1, 5 and 14). While we definitely agree that creativity will be a fundamental asset for well-being in the future (Corazza 2017), we also need to recognize that thinking and acting creatively is not a panacea for all situations and all contexts. Creative behavior should always coexist with intelligent behavior, and striking the optimal balance in the development of these two fundamental constructs should be the real target of the design of new educational pedagogies (Corazza and Lubart 2021; Corazza et al. 2021a, b).

Ingunn Johanne Ness There seems to be a consensus that technology and all its benefits are very important when it comes to education in the Twenty-First Century. Still, one could argue that the focus should not be on technology alone but also on pedagogy – a pedagogy that will equip and empower students for the twenty-first century. Both dialogic pedagogy and education in the twenty-first century have in common that they want to empower learners by developing competencies that go beyond content knowledge (Teo 2019).

In this section I want to make a point about both how technology needs to be a part of the whole pedagogical design and that it is important that teachers not only put an emphasis on what their students are going to learn, but also teach them how to learn. Both have pedagogical implications.

Most scholars agree that one of the most important tasks of education is to prepare learners for the future. This is a big task which involves reflecting on future competency requirements (Ness 2020). Students will need to adapt to a work life and solve complex challenges that are largely unknown to us at this point. The only thing we can be certain of is that solving future challenges will require new forms of (digital) literacies, creative problem-solving skills, collaboration and communication skills, cultural and ethical awareness as well as lifelong learning. Such skills are often referred to as part of Twenty-First Century skills and have in common that it is about going beyond the learning of content knowledge and instead focusses on a more holistic education that emphasizes life skills that are relevant and necessary to cope in the twenty-first century workplace (Beers 2011). In other words, it is no longer enough to teach students what to learn; we also need to teach them how to learn (Teo 2019).

This is also of interest for many educational researchers in the interdisciplinary field of Creativity, Learning and Technology. The field has received a lot of attention in the last couple of years, in particular due to the COVID-19 pandemic. There was an overnight need for a digital transformation to support teaching and learning. All communication was forced to move online, and teachers had to start using different technologies that for many of them, were new. In this way the crisis also influenced their teaching designs, created a need to develop new practices involving different digital tools, and in the end challenged their pedagogical approaches. At the same time as any crisis will be experienced as challenging (Beghetto 2021), there is also a large creative and innovative potential involved and the COVID-19 pandemic can be seen as a catalyst of the need to innovate the process of teaching. For many teachers it was their first venture into online teaching, and many were stressed as a result of the uncertain situation and finding themselves outside their comfort zone. The experience of uncertainty is of course uncomfortable, but it is also an opportunity to think and act in new ways (Chap. 14). No doubt, some teachers found this new situation very difficult and offered their students no more than video-conferenced lectures or voice-over powerpoints and email answers to questions. This approach became quite monological, and some students reported that they missed dialogues and the social interaction with teachers and other students and that they learned less when they could not discuss with their peers. Other teachers were more adventurous and took a variety of new tools and pedagogical practices into use in order to still create social learning spaces in their digital classroom. Video conferencing tools, like Teams or ZOOM were examples of teaching tools that made it possible for teachers and students to communicate. In ZOOM it was also possible to discuss in smaller groups (break out rooms) and in plenum. It soon became clear that teachers displayed a very different use of technological tools, also including which tools they made use of. Technology can support teaching approaches that open up for creative learning (Chap. 14) which is about how educators design learning experiences with the goal of promoting new and meaningful learning in and also beyond the immediate learning event – but then the technology needs to be a part of the whole pedagogical learning design. What is important is that all teachers have access to and can use technology that can support their individual teaching needs that can go beyond teaching students' basic literacy to also involving the different necessary skills for the future, – meaning teachers need to reflect on both *what* and *how* they want their students to learn. It is therefore crucial to focus on the pedagogy behind the teachers' didactical choices since this will influence the different ways teachers make use of technology.

Dialogic pedagogy is a theoretical framework with an emphasis on exploring open questions, listening, arguing and thinking together in an equal dialogue. – The aim is to broaden our understanding (Alexander 2008). However, it is important to note that dialogical teaching is not a method, it is instead a basic quality that should be present in all teaching, no matter the subject. Curiosity, doubt, questions, respect, critical exploration, reflection – these are all important aspects in dialogical teaching and might support students' ability to live in an age of uncertainty and to use this as a benefit (Chap. 14). Students learn how to learn and to engage in dialogue with different voices and perspectives. This polyphony can be used as a creative stimulus and open up for new understandings (Ness and Riese 2015; Ness 2021). Technological tools can very well support such dialogues in the classroom if they are used as a part of such a pedagogical learning design. This is why we have to focus not only on the technologies that are becoming more and more advanced but also on the pedagogy that can help us to achieve a sustainable future.

Vlad Glăveanu Corazza and Ness's commentaries elevate this discussion to what we can call the bigger picture around creativity, learning and technology. They both, in different ways, tackle the key questions of this emerging field and, like most questions, these revolve around the 'who', 'what', 'how' and 'why' of the phenomena we are discussing about. I am particularly persuaded by the idea that we should move away from focusing only on 'how' technology relates to, informs, enhances or, at times, inhibits creative learning. While interrogations about process remain vital within fields that consistently prioritized product and person (especially in creativity studies; Barron and Harrington 1981), we need to be able to connect processes with the actors, actions, audiences, artefacts, and affordances they are embedded within (Glăveanu 2013). In other words, we should reflect on 'how' we engage technology in view of 'who' is involved in this engagement, 'what' artefacts and resources are being mobilized by it, and 'why' or 'for what' educational purposes.

Corazza's notion of 'cyber-humans' seen as an integrated whole instead of two 'elements' – technology and human beings – coming together in ways that can be coupled and de-coupled from each other, is very appealing from a sociocultural perspective. In this approach we have been introduced since Vygotsky's seminal work (see Van der Veer and Valsiner 1994) to the notion that people, signs and tools are systemically connected with each other and, more than this, that the human mind (and forms of action) would be impossible without cultural and technological mediation. The advance in digital technologies and the new context emerging from the pandemic have merely accelerated what was already our ontological condition as psychological, technological, and embodied beings. This non-dualist understanding has deep implications for how we answer the 'who', 'what', 'how' and 'why' questions raised throughout this book. Instead of talking about whether technology is facilitating or not learning and creativity we should be asking differentiated questions about the ways in which technology-use shapes people, objects, places, practices, and power relations within society.

Ness reminds us of the importance of the broader epistemological framework we are working within. Instead of an old transmission model of knowledge – or what Freire (1970) called the 'banking model of education' – in which teachers deposit information in the minds of their students, we are consistently moving towards the wide scale use of dialogical, inquiry-based pedagogies. Acknowledging the integral role played by technology in these educational developments is not enough. As Ness mentions, we need to connect 'what' and 'how' students create and learn with technology. We need to open up dialogical spaces in the classroom, including about education, new technologies, and what it means to live 'cyber-human' lives. Dialogism here becomes not only a method or a means to an end but part of the end or the goal itself. Critical reflection and creative engagement should permeate education, independent of its level and discipline, instead of being added on to the curriculum in designated (but contained) classes and exercises. The 'where' and 'when' of these practices (besides the 'who', 'what', 'how' and 'why') become important for developing students, learners and creators adapted to this turbulent twenty-first century.

Ingunn Johanne Ness As Glăveanu says, critical reflection and creative engagement should permeate education, independent of its level and discipline. I believe that it is not enough to teach students creative engagement, creative learning or creative problem solving as a specific technique. The weakness of such a way of thinking is that specific methods will be too narrow, and it is also not sufficient to link creativity to specific disciplines and over short periods of time. Corazza says that while we agree that creativity will be a fundamental asset for well-being in the future, we need to recognize that thinking and acting creatively is not the only solution for all situations and all contexts and always need to be balanced with intelligent behavior. This is also something that needs to be reflected upon when designing new educational practices.

I would also argue that when designing new educational pedagogies, both critical and creative thinking must be balanced and seen in the context of our greater global society. Thus, the questions on the 'where' and 'when' of these practices, in addition to the 'who', 'what', 'how' and 'why', become important for developing learners adapted to this turbulent twenty-first century, as Glăveanu states. Neither critical thinking or creative processes are specific, instead one must think more holistically and let both critical thinking and creativity be part of the whole curriculum in schools and take place across courses at university level. It is therefore important that we can offer arenas where pupils and students can practice creative processes through dialogue with open and exploratory questions in all subjects and cross subject boundaries over time so that this becomes something that has continuity in the pupils 'and students' everyday lives. This is demanding for any teacher. It is also a challenge when teachers are faced with an education system with testing and a results-oriented approach that does not primarily allow for creativity and exploration of different perspectives which in turn can hinder valuable creativity in the classroom.

Dialogue is the central concept in the works of Mikhail Bakhtin, a Russian philosopher of language, literature and culture. Bakhtin's view on dialogue comprises both an ontological and an epistemological understanding. Dialogue was, according to Bakhtin (1981), both a fact of life and at the same time an ideal to strive for. True dialogues have a plurality of perspectives, Therefore dialogism is so important in order to avoid echo-chambers with dangerous belief systems that are not challenged., Freire's work on the role of dialogue in the empowerment of 'the oppressed' (1970) shows how dialogue is crucial in true education. Dialogue enables us to 'name' the world and claims that this dialogue cannot be reduced to the act of one person's "depositing" ideas in another (Maine and Wegerif 2022).

Giovanni Emanuele Corazza I agree with both Ness and Glăveanu, whose comments help us grasp the complexity of the issue we are trying to tackle, which cannot be addressed with simple, ready-made solutions. In fact, the critical reflection and creative engagement that are called for by both Glăveanu and Ness should be applied not only in the classroom but also in the meta-level of designing the education system. But anyone who has tried to propose reforms of the education system that go beyond an educational status quo, which finds its roots back in the first industrial revolution, knows that resistance to this specific change is extremely strong.

It is interesting to try to briefly analyze the nature of this resistance, as well as its original cause. Since the eighteenth century, i.e., the start of the industrial revolution, the strongest socio-economic trend became that of *standardization* (Corazza 2016; Corazza et al. 2010). This is because an economy based on industry implies vast productions of standard products and massification of taste, in order for those products to be sold and consumed. Somewhat inevitably, the trend for standardization percolated also to the educational realm: the majority of the population was illiterate and not ready for the lifestyle of the industrial society, and it was necessary to raise the level of knowledge and competence of the masses. Standardized education and testing became more and more established, up until today. Whereas this had clear positive sides in terms of raising the general cultural level, it also created a system which tends to preserve itself: the strength of any standard is directly proportional to the number of people/institutions adopting it, and educational standards are no exception.

Projecting the above observations onto the topics discussed in this book, a key question can be raised about whether the available new technologies will help moving away from the above standardization trend in education, opening up a new era for learning and creating. Interestingly, the answer is far from obvious, as both positive and negative elements can be identified.

On the positive side, new technologies allow anyone to access information on any knowledge domain, thus enabling interdisciplinary and transdisciplinary approaches that would be very difficult otherwise (Corazza et al. 2021a, b). This points to the possibility for a sort of new Renaissance, an epoch in which anyone will have the chance to go well beyond the narrow limits of a single discipline and a single profession. Also, personalized learning paths can be designed through AI-enabled systems, leading possibly to the fuller realization of everyone's potential. On the other hand, there might also be negative sides, potentially quite dangerous.

First of all, learning through virtual classes might start a trend towards homogeneity of knowledge, possibly caused by the fact that certain educational models will win over others in the battlefield of the global competition of education in virtual worlds. It would certainly be in the interest of superpowers to impose their educational model even beyond the geographical borders of their political control. More subtle but also more dangerous is the possible control of both search engines and data-based AI algorithms, which have already become but will more and more be the sources of information through which we will answer most of our questions. Biasing the response of an Internet search engine or the outcome of a big-data analysis might be powerful instruments for hacking culture. As touched upon by Glăveanu above, to prevent or counter these negative sides, ethics will be crucial, playing a major role in ensuring a new form of human rights, which we could define as 'equitable and unbiased access to digital information in educational, professional, and everyday settings'.

Vlad Glăveanu It is always exciting to explore both the challenges and opportunities of new or at least newly introduced technologies in educational spaces. As Corazza rightfully notes, inter- and transdisciplinarity can be immensely helped by increased connectivity; at the same time, issues of power and monopoly over technology itself, its algorithms, and the pedagogical uses of both, are starting to come to the fore. Ness comes back to the topic of dialogue that seems to be a recurrent theme, at least within sociocultural approaches to creativity and learning. Educational technologies are and will continue to massively impact how dialogues unfold, from the range of perspectives they invite to the processes through which they build consensus or lead to further divergence. The medium might not be the same as the message, yet it certainly contributes to how we receive and interpret messages (and understand who is sending them) in both education and beyond.

One interesting question about technology and dialogue, however, is whether we can truly create dialogical spaces within digital environments, either in the form of human-computer or computer mediated interactions with other fellow humans. Despite a resurgence, in recent years, of post-humanist thought (Wolfe 2010), advancing the view that human and non-human actors share agency and engage in equal terms in assemblages we refer to as institutions and social practices, critics are quick to point to the loss of authenticity in online human relations. Some authors more or less explicitly question whether technology is indeed conducive for dialogicality (e.g., Marková 2017; Jovchelovitch 2019) or rather controlling and, when overused, alienating. In what ways can we foster authentic dialogues, in education, in spaces that miss the embodied presence of the other and, frequently enough, present us with stereotypical views of who the other is?

Giovanni Emanuele Corazza Reacting to the question posed by Glăveanu about the possibility to truly create dialogical spaces within digital environments, I believe we need to distinguish two fundamental elements: the medium and the actors. Regarding the medium, it can be argued that no digital experience will ever be able to completely match the experience of a physical dialogue between humans, in which communication exploits several parallel strands going well beyond the spoken words. Perhaps, when the on-going development of technology for fully immersive experiences (able to carry more that audio and video) will have reached maturity, we will have a digital experience that is as close as possible to a physical one; but I feel something will always be missing.

On the other hand, reduced forms of communication that exploit sensorially limited services, such as videoconferencing, video calls, voice calls, texting, or email cannot be ruled out as supports for creating effective, albeit different, forms of dialogue. In fact, the limitations of the medium can even be turned to advantage if one is able to master its characteristics. To make a very significant example that does not relate to the digital world, in prehistoric times all that humanity possessed for transferring information was face-to-face narration, i.e. physical dialogues. When the technology of "writing" was introduced in early civilization, a totally new world of virtual (in the sense of indirect) dialogues was opened, certainly with different characteristics, but nonetheless extremely powerful in generating exchanges across cultures and time epochs. In essence, we believe that all digital communication media are helpful in supporting dialogues, although in a variety of modalities and with diversified effectiveness.

Regarding a dialogue between human actors, it is interesting to note that their dialogical ability, as well as their personality, can vary considerably across communication experiences over different media. In other words, it appears that today a person has no "single" personality or dialogical ability, but several of these, depending on the diverse communication supports one uses. An important problem for this actor would therefore be to maintain coherence amongst these different "selves", while knowing at the same time that what one shows on a general-purpose social network will never be the same as what appears on a chat with close friends, or on a professional Internet blog.

Finally, the most intriguing case might be that of a dialogue on a digital communication medium in which the involved actors include also (or only) artificially intelligent machines: could this still be an *authentic* dialogue? This question definitely calls for a philosophical analysis which cannot be attempted in a few lines. Let it suffice here to say that we believe authenticity in this case is not attainable today but perhaps tomorrow, as it will require to endow machines with awareness, i.e. with conscience. Whereas this might seem to be an impossible task today, work is under way to give a mathematical basis to the seemingly unseizable concept of conscience, opening up the possibility for its artificial production (Tegmark 2015; Tononi 2012).

Ingunn Johanne Ness New technology offers both opportunities and challenges and I agree with both Corazza and Glăveanu in having a critical mind to both. Glăveanu brings up a point regarding technology and dialogue and whether we can truly create dialogical spaces within digital environments. Corazza leads us to reflect on whether a dialogue on a digital communication medium in which the involved actors include artificially intelligent machines - could this still be an *authentic* dialogue? No doubt, these are interesting and important questions to address.

When taking a dialogic approach to education with ICT, this includes a variety of tools, and artificially intelligent machines are included in such tools. According to Wegerif and Major (2019) AI makes us consider how we can incorporate the alienness of this technology into an expanded dialogue – in the end, we must consider how to expand what it means to be human, to be us. In addition to being a philosophical question, this is also an educational challenge.

Dialogic education can open up and broaden dialogic spaces within digital environments through the potential in bringing students across cultural and geographic differences together using the internet for instance, and also scaffold dialogues using synchronous and asynchronous environments (Wegerif 2007). Still, the question still remains, can we foster true dialogues in spaces when we actually also miss the embodied presence of the other person and in some programs, present a stereotypical view of who the other is?

Vlad Glăveanu These are both very thoughtful considerations of this 'big' question of whether technology use can indeed be dialogical. As Corazza notes, this is not a question only for today but also tomorrow. Although I personally am highly skeptical about machines reaching human types of consciousness (perhaps it will be a different category altogether), I do appreciate that both old and new technologies, for as primitive or as sophisticated as they are, have always participated within and supported dialogue and, with it, the creative emergence of meaning. Ness, building also on Wegerif and Major (2019), invites us to take this opportunity and reflect on what it means to be human in ways that expand our self-understanding. These are not opportunities to be missed, especially in education.

My personal view about this issue is that *both* human-human and human-computer interactions can be dialogical or non-dialogical (monological). The distinguishing criterion, following a Bakhtin approach, is whether tension, emergence and novelty are embedded within the exchange. We can all certainly think of moments of communication with fellow human beings that are not described by any of these or, at the very least, these features are not dominant. In education, this includes interactions that are based purely on the transmission of information, normally from teacher to student, with no 'creative openings' or room left for wonder and uncertainty (Gajda et al. 2017; Glăveanu 2020b). In contrast, the use of technology either as a tool for communication with others or as an end in and of itself, can trigger moments of questioning, reflection and even awe (e.g., Chirico et al. 2018). The quality of 'dialogical', for me, is pragmatically in the action and not decided *a priori* by the medium, actors, or message.

And with this we are close to bringing our own dialogue to an end. But, before we do, final reflections are in order. What has our own dialogue here about learning, creativity and technology (a dialogue that, by the way, is heavily mediated by technology!) teach us? What kind of surprises, tensions and questions emerged with its help? What did we each learn from it?

Ingunn Johanne Ness Glăveanu indeed raises an interesting question above, what did this dialogue mediated by technology teach us? Well, technology made it possible for us to engage in this dialogue and it has been very interesting to exchange thoughts throughout this chapter and to co-create a chapter together. Co-creation leads to understanding and according to Bakhtin (1986) this is something that emerges in relational dialogues as a response to utterances and I want to include a quote by Bakhtin:

> The word in language is half someone else's. It becomes one's "own" only when the speaker populates it with his own intentions, his own accent, when he appropriates the word, adapting it to his own semantic and expressive intention. Prior to this moment of appropriation, the word does not exist in a neutral and impersonal language... but rather it exists in other people's mouths, in other people's contexts, serving other people's intentions; it is from there that one must take the word, and make it one's own (Bakhtin 1981, p. 294).

According to Bakhtin it is through meetings with other perspectives and utterances, new knowledge occurs and it is in our seeking to understand the alien, that we can learn something new. This is about creative understanding, according to Bakhtin (1986). To him, creativity and understanding is closely linked together. So, in this case, writing together and co-creating this chapter, the use of technology has worked as a tool for communication and triggered moments of questioning, reflection and new understandings. I have truly been inspired by Corazza and Glăveanu - and I agree with Glăveanu, that the quality of 'dialogical', is in the action and not decided *a priori* by the medium, actors, or message. In the end, maybe dialogue should be seen as an end itself . . .

Giovanni Emanuele Corazza I like taking pictures and sharing them. Perhaps this is one of the most positive affordances of social networks and chats, when used wisely. There is no equivalent experience to this instantaneous form of shared visual communication without the supporting technology developed in the last 70 years in telecommunications and computer science. But as I was reflecting on Glăveanu's final questions, it dawned on me that this multi-handed writing of a chapter could be metaphorically interpreted as the flash shot of a dialogue, wherein our opinions and exchanges become fixed on paper as they were expressed in the moment! And I imagine both Ness and Glăveanu smiling as their fingers were frantically dancing on the keyboard…

An interesting idea, taking pictures of a dialogue… or was this rather a trialogue, as there are three of us involved? Actually, the word 'trialogue' was invented in 1532 by no one less than St. Sir Thomas More (Swidler 1994), but it was later considered a misinterpretation of the etymology of dialogue, a mistake caused by breaking it down as di(*two*)-a-logue(*word*) instead of the correct dia(*across*)-logue(*word*). In other words, a dialogue can involve two or more people. However, we shouldn't throw away the provocation of the trialogue too fast. In fact, anyone involved in creativity studies knows that 'errors' are actually the door to ideas that have a potential for great originality and effectiveness, perhaps emerging only a-posteriori (Corazza 2016)! Let's take seriously the strong push towards tri-partite entities that is carried by Thomas More's trialogue.

Indeed, we live in a world dominated by dualities and dichotomies, analyses and cultural disputes are almost always dominated by two opposing sides. It would therefore seem that human understanding is greatly helped by working in between two extremes, in a bi-dimensional world. Given that our physical reality is three-dimensional, it is a very reasonable simplification, but also a clear reduction. How would our cultural discourse evolve if we were to systematically introduce 'trialities' and 'trichotomies' as fundamental frameworks for understanding? A question that is best left unanswered until our next trialogue with Ness and Glăveanu!

Vlad Glăveanu Our trialogue is coming to an end but, in the good tradition of dialogical creativity (Ness and Glăveanu 2019), the end is merely the beginning of new processes of meaning-making, reflection, and continued action. The metaphor – and reality – of the dialogue applies to this entire volume. As an edited book, it already includes a polyphony of voices, perspectives, and author backgrounds. More than this, as each chapter focuses on provocations, they necessarily have to cross temporal, conceptual, and disciplinary boundaries. The 'inner' dialogicality of every contribution is enhanced by the fact that you, as a reader, can decide on your own path through the book, including any detours and back and forth movements you want to make. In this way, your own reading will recreate the space of possibility this volume aims to inspire as it imagines new futures for learning, creativity, and technology.

Our 'task' here was to illustrate how an open-ended, playful dialogue taking its point of departure from the contributions included within these covers can lead to new questions, questions to which you surely have answers of your own. In many ways, this last chapter is meant to be a new opening, rather than closing, of the range of creative provocations showcased here. Some of these dilemmas have been around for a long while, others we are beginning to discover today, prompted by a unique societal and global context. The story of the links between creativity, learning and technology is certainly very old as it arguably goes back to our emergence as a tool-using, tool-transmitting, and tool-inventing species (Festinger 1983). But what we learn and how we create with technology remain key questions that will only become more visible and more pressing over the next years and decades. And so the dialogues, debates and provocations around these themes will also intensify. We are grateful for the opportunity to contribute to a book that marks an important moment in this evolution.

Here's to the future and to the next opportunity to engage in creative trialogues about it!

References

Alexander, R. J. (2008). *Towards Dialogic Teaching: Rethinking Classroom Talk.* Fourth edition. York: Dialogos.

Bakhtin, M. M. (1981). *The Dialogic Imagination: Four Essays.* Austin: University of Texas Press.

Barron, F., & Harrington, D. M. (1981). Creativity, intelligence, and personality. *Annual Review of Psychology, 32*(1), 439–476.

Bakhtin, M. (1986). *Speech genres and other late essays* (Vol. No. 8, University of Texas Press Slavic series). Austin, TX: University of Texas Press.

Beers, S. (2011). *Teaching 21st century skills: An ASCD action tool.* Virginia: Association of Supervision and Curriculum Development.

Beghetto, R. (2021). How Times of Crisis Serve as a Catalyst for Creative Action: An AgenticPerspective. *Frontiers in Psychology*. 11. 600–685. https://doi.org/10.3389/fpsyg.2020.600685.

Chirico, A., Ferrise, F., Cordella, L., & Gaggioli, A. (2018). Designing awe in virtual reality: An experimental study. *Frontiers in Psychology, 8*, 2351.

Clark, A., & Chalmers, D. (1998). The extended mind. *Analysis, 58*(1), 7–19.

Coin, A., & Dubljević, V. (2021). The authenticity of machine-augmented human intelligence: Therapy, enhancement, and the extended mind. *Neuroethics, 14*(2), 283–290.

Copeland, S. (2022). Ethics of Possibility. In: The Palgrave Encyclopedia of the Possible. Palgrave Macmillan, Cham. https://doi.org/10.1007/978-3-319-98390-5_263-1.

Corazza, G.E. (2016). Potential originality and effectiveness: The dynamic definition of creativity, *Creativity Research Journal. 28*(3), 258–267.

Corazza G.E. (2017). Organic creativity for well-being in the post-information society. *Europe's Journal Of Psychology, 13*, 599–605.

Corazza, G. E., and Lubart, T. (2021). Intelligence and Creativity: Mapping Constructs on the Space-Time Continuum. *Journal of Intelligence 9: 1.* https://doi.org/10.3390/jintelligence9010001

Corazza, G. E., Vanelli-Coralli, A., & Pedone, R. (2010). Technology as a need: Trends in the evolving information society. *Advances in Electronics and Telecommunications, 1*(1), 124–132.

Corazza, G. E., Darbellay, F., Lubart, T., & Panciroli, C. (2021a). Developing Intelligence and Creativity in Education: Insights from the Space–Time Continuum. In *Creativity and Learning* (pp. 69–87). Palgrave Macmillan, Cham.

Corazza, G. E., Reiter-Palmon, R., Beghetto, R., & Lubart, T. (2021b). Intelligence and Creativity in the Space-Time Continuum for Education, Business, and Development. *Journal of Creativity*, 100003. https://doi.org/10.1016/j.yjoc.2021.100003

Festinger, L. (1983). *The human legacy.* New York, NY: Columbia University Press.

Freire, P. (1970). *Pedagogy of the oppressed.* New York, NY: Continuum/Seabody.

Gajda, A., Beghetto, R. A., & Karwowski, M. (2017). Exploring creative learning in the classroom: A multi-method approach. *Thinking Skills and Creativity, 24*, 250–267.
Glăveanu, V. P. (2013). Rewriting the language of creativity: The Five A's framework. *Review of General Psychology, 17*(1), 69–81.
Glăveanu, V. P. (2020a). *The possible: A sociocultural theory*. New York, NY: Oxford University Press.
Glăveanu, V. P. (2020b). *Wonder: The extraordinary power of an ordinary experience*. London: Bloomsbury Press.
Hartley, L. P. (1953). *The Go-Between*. Harmondsworth: Penguin.
Jovchelovitch, S. (2019). *Knowledge in context: Representations, community and culture*. London: Routledge.
Maine, F., Wegerif, R. (2022). Dialogism. In: The Palgrave Encyclopedia of the Possible. Palgrave Macmillan, Cham. https://doi.org/10.1007/978-3-319-98390-5_239-1
Marková, I. (2017). From imagination to well-controlled images: Challenge for the dialogical mind. In T. Zittoun & V. P. Glăveanu (Eds.), *Oxford Handbook of Culture and Imagination* (pp. 315–349). Oxford: Oxford University Press.
Ness, I. J. (2020). Polyfoni og kreative vidensprocesser i klasserummet og i videregående uddannelser. In: O. Dysthe, I. J. Ness & P. O. Kierkegaard (Eds), *Dialogisk pædagogik, kreativitet og læring* (pp. 113–143). Aarhus: Forlaget Klim
Ness, I.J. (2021). Polyphony. In V. P. Glăveanu (Ed.), *The Palgrave Encyclopedia of the Possible*. London: Palgrave.
Ness, I. J., & Riese, H. (2015). Openness, curiosity and respect: Underlying conditions for developing innovative knowledge and ideas between disciplines. *Learning Culture and Social Interaction, 6*(September 2015), 29–39.
Ness, J. I., & Glăveanu, V. P. (2019). Polyphonic orchestration: The dialogical nature of creativity. In Beghetto R. & Corazza G. (Eds), *Dynamic Perspectives on Creativity. Creativity Theory and Action in Education* (pp. 189–206). Cham: Springer
Swidler, L. (1994). Trialogue? Absolutely!. *CrossCurrents*, 140–143.
Tegmark, M. (2015). Consciousness as a state of matter. *Chaos, Solitons & Fractals, 76*, 238–270.
Teo, P. (2019) Teaching for the 21st century: A case for dialogic pedagogy. *Learning Culture and Social Interaction,* 21, 170–178.
Tononi, G. (2012). *Phi: A Voyage from the Brain to the Soul*. New York: Pantheon.
Van der Veer, R., & Valsiner, J. (Eds.) (1994). *The Vygotsky Reader*. Cambridge, MA: Blackwell.
Wegerif, R. (2007). *Dialogic Education and Technology: Expanding the space of learning*. New York: Springer-Verlag.
Wegerif, R. & Major, L. (2019). Buber, Educational technology, and the expansion of dialogic space. *AI & SOCIETY, 34*, 109–119.
Wolfe, C. (2010). *What is posthumanism*? Minnesota, MN: University of Minnesota Press.

Vlad Petre Glăveanu, PhD, is Full Professor of Psychology in the School of Psychology at Dublin City University, and Professor II at the Centre for the Science of Learning and Technology (SLATE), University of Bergen. His work focuses on creativity, imagination, wonder, culture, and human possibility. He is author of The Possible: A Sociocultural Theory (2020) and Creativity: A Very Short Introduction (2020), with Oxford University Press. He is the founder and president of the Possibility Studies Network and editor of the Sage journal Possibility Studies and Society, as well as the upcoming book series Cambridge Series on Possibility Studies. He received in 2018 the Berlyne Award from the American Psychological Association for outstanding early career contributions to the field of aesthetics, creativity, and the arts.

Giovanni Emanuele Corazza, PhD, is a Full Professor at the Alma Mater Studiorum University of Bologna, President of the Fondazione Guglielmo Marconi, founder of the Marconi Institute for Creativity, Member of the Marconi Society Board of Directors. His research interests are focused on the development of the Dynamic Creativity Framework. He was Member of the Partnership Board of the 5G Infrastructure Association in the years 2013–2018, Member of the Board of Directors of the University of Bologna in the years 2012–2018, President of the CINECA consortium for supercomputing in the years 2017–2019, Head of the Department of Electronics, Computer Science and Systems (DEIS) in the years 2009–2012, Chairman of the School for Telecommunications in the years 2000–2003, Chairman of the Advanced Satellite Mobile Systems Task Force (ASMS TF), Founder and Chairman of the Integral Satcom Initiative (ISI), a European Technology Platform devoted to Satellite Communications.

Ingunn Johanne Ness, PhD, is a Senior Researcher and leader of the theme Creativity, Learning & Technology at the Centre for the Science of Learning & Technology (SLATE), University of Bergen. She is an expert on learning and creativity in the field of education and business with a particular interest for dialogical pedagogy and the sociocultural approach. She works with one of the world's leading environments on sociocultural theory, the OSAT group at the University of Oxford and has close collaboration with Webster Center for Creativity and Innovation in Geneva. She is a member of the International study of Creativity and Innovation (ISSCI) and has a number of publications in International journals and Handbooks.

The manufacturer's authorised representative in the EU is Springer Nature Customer Service Centre GmbH, Europaplatz 3, 69115 Heidelberg, Germany. If you have any concerns regarding our products, please contact ProductSafety@springernature.com

Printed and bound by CPI Group (UK) Ltd, Croydon, CR0 4YY

28/09/2026

02224077 0002